Incentives and Instruments for Sustainable Irrigation

Incentives and Instruments for Sustainable Irrigation

Edited by

H. Bjornlund

University of South Australia, Australia
and
University of Lethbridge, Canada

WIT*PRESS* Southampton, Boston

Editor:
H. Bjornlund
University of South Australia, Australia and *University of Lethbridge, Canada*

Published by

WIT Press
Ashurst Lodge, Ashurst, Southampton, SO40 7AA, UK
Tel: 44 (0) 238 029 3223; Fax: 44 (0) 238 029 2853
E-Mail: witpress@witpress.com
http://www.witpress.com

For USA, Canada and Mexico

WIT Press
25 Bridge Street, Billerica, MA 01821, USA
Tel: 978 667 5841; Fax: 978 667 7582
E-Mail: infousa@witpress.com
http://www.witpress.com

British Library Cataloguing-in-Publication Data

A Catalogue record for this book is available
from the British Library

ISBN: 978-1-84564-406-2

Library of Congress Catalog Card Number: 2009930794

The texts of the papers in this volume were set individually by the authors or under their supervision.

Printed in Great Britain by MPG Books Group, Bodmin and King's Lynn.

Contents

Part I: Concepts and Background

Chapter 1

Sustainable irrigation: Setting the scene

Henning Bjornlund
University of South Australia and University of Lethbridge, Alberta, Canada

1 Introduction

Including the words "Sustainable Irrigation" in the title of a book necessarily raises the question *what is sustainable irrigation?* The terms sustainability and sustainable development have entered our language over the past 30 years and have gained increased use, often leaving the impression that they have become fashionable "buzzwords" used by politicians and companies because it is the "right" thing to say to increase votes or sales. The terms have also found their way into modern legislation, usually in the context of preambles or objectives, rather than in operational clauses. They are therefore less likely to have an impact on outcomes and can give rise to confusion and litigation.

This book will not weigh into the debate over definitions of sustainability or sustainable development or the difference between the two. That literature is already waste with an overwhelming array of often contradictory definitions (Van pelt *et al.*, 1992; Flint, 2004). The purpose of this book is to discuss incentives and instruments which can be used to promote sustainable irrigation or assist irrigation to stay within sustainable limits as discussed in this chapter.

Sustainable development is a kind of overarching or parent concept which among its offspring has sustainable agriculture and sustainable water management; sustainable irrigation is a lower order concept relative to these two. Sustainable irrigation cannot exist without sustainable agriculture and sustainable water management. Most discussions of sustainable development and lower-order concepts make reference to the Brundtland Report *Our Common Future* (WCED, 1987), which refers to sustainable development as development that meets the needs of the present without compromising the ability of future generations to meet their own needs. The ability to achieve this is closely linked to the state of the resource base (Csurgo *et al.*, 2008); in the case of sustainable water management, a core issue is healthy ecosystems. Without healthy ecosystems, water resources cannot continue to meet a constant or growing level of needs into the future. The "needs" of current and future generations can also not be dealt with in isolation; but they have to be seen in the context of a certain watershed and the social, economic and environmental aspirations of its stakeholders and within a framework of higher order priorities as expressed in state or national water plans. Viewed at the national level, it might be desirable to define healthy ecosystems and environments differently for different water sources as some have already been worked so hard that parts of ecosystems have been permanently destroyed. For example, in Australia a distinction is made between working rivers and wild rivers with very different interpretation of what a healthy river looks like, and hence what constitutes a healthy ecosystem.

From this discussion, it is apparent that sustainable water management must

1. be the joint responsibility of governments, communities and water users such as farmers (Blomquist *et al.*, 2005);
2. be seen in the context of a specific location, at a specific period in time, while acknowledging that it is subject to dynamic environmental and socio-economic forces requiring flexibility and adaptive management (Brierley *et al.*, 2006);
3. be seen in the context of a specific set of stakeholder interests (Brierley *et al.*, 2006) and
4. reflect the local environmental, social, economic and cultural values (Elliott, 2004). It has been argued that society must determine to what extent ecosystem needs will be met in a particular watershed (Gleick, 1998a.; Postel & Richter, 2003).

Hence, sustainable water management involves a compromise that balances environmental, social and economic needs and is acceptable to the parties involved; the watershed is therefore the most appropriate level for water planning within the context of subsidiarity (Walmsley, 2002; Henriksen & Barlebo, 2008; Jordan & Jeppeson, 2000). Devolving the decision-making processes of water management to the watershed level and involving local stakeholder groups have therefore been central parts of both national and international water policies since Agenda 21, the program for action for sustainable development (UNCED, 1992).

Sustainable agriculture includes the use of agricultural practices, including irrigation techniques and on-farm water management practices, which enable current land and water resources to meet the needs of current and future generations. To do this, such techniques and practices cannot degrade the on-farm resource base or the resource system supporting it. The United States Congress defines sustainable agriculture as "... an integrated system of plant and animal production practices having a site-specific application that will, over the long-term: (1) satisfy human food and fiber needs; (2) enhance environmental quality and the natural resource base upon which the agriculture economy depends; (3) make the most efficient use of non-renewable resources and on-farm resources and integrate, where appropriate, natural biological cycles and controls; (4) sustain the economic viability of farm operations; and (5) enhance the quality of life for farmers and society as a whole (USDA, 2008)".

2 Sustainable irrigation in the context of this book

In the context of sustainable irrigation, it has to be understood that irrigation in any specific location depends on a water resource system which yields an annual flow of water of a certain quality. Sustainable irrigation requires that extraction for consumptive use is kept at a level where sufficient water is left in the resource to ensure an aquatic ecosystem which can maintain a level of water quality sufficient for crop growth. This level of extraction might be significantly lower than what other stakeholders require to satisfy their needs. Therefore, stakeholders must first define sustainable water use within their watershed, that is,

1. define the acceptable level of water quality, thereby defining which environment or ecosystem they want to sustain and therefore how much water will need to be left in the river to meet in-stream flow needs in the form of base flow as well as to create environmental events such as flooding of wetlands (Postel & Richter, 2003);
2. define other required cultural, social and public good outcomes such as setting water aside for minority groups or previously disadvantaged groups as in South Africa (Backenberger, 2005);
3. establish how much water the system yields;
4. define the pool of water which can be sustainably used for extractive purposes by subtracting one and two from three above. In Australia, this is called the consumptive pool;
5. allocate the consumptive pool among competing users; and
6. acknowledge that sustainable water and land management cannot be separated and both have to be dealt with as part of an integrated watershed management approach (Blomquist *et al.*, 2005). Hence, the foundation for sustainable agriculture is also defined as part of the planning process.

These are issues which are most appropriately dealt with by governance structures (de Loe and Bjornlund, Chapter 3). Issues such as conflict resolution, consensus building, water allocation and defining community, cultural and environmental needs have to be dealt with in adaptive watershed planning processes. These, in turn, have to be seen in the context of basin, state and/or federal policies/plans defining sustainable water management at the national level. These processes might include tradeoffs between wild rivers and working rivers to achieve national- or state-level environmental objectives.

Once the consumptive pool has been defined, the issue of sustainable irrigation can be considered. Sustainable irrigation has at least two main aspects: (1) irrigation must be supported by sustainable irrigation development which is capable of delivering a constant and reliable supply of sufficient-quality water for irrigation to support a constant or increasing level of food production while keeping its impact on the environment within sustainable levels as defined in the watershed plans (Easter and McCann, Chapter 3) and (2) individual irrigators must adopt land and water management practices which do not degrade the land resource and keep their impact on the river and groundwater environment within acceptable levels while allowing a constant or increasing level of food production. This includes a process of adopting more efficient and productive irrigation technology and water management practices as they are developed (Nicol *et al.*, Chapter 12).

In many water-stressed areas of the world, the consumptive pool is likely to be smaller than the pool of water which has previously been used by consumptive users. This is particularly likely to be the case with respect to irrigators' share. Under scarcity, the planning process is likely to give higher priority to human needs for drinking water and sanitation and to the needs of ecosystems than the needs of irrigation (Gleick, 1998b; Postel & Richter, 2003). The challenge then becomes "how can irrigation maintain its production output with access to less water while preventing negative environmental impacts".

If we look at "irrigation" in a broader sense, by also including what we could term the irrigation communities whose existence is dependent on irrigation for continued viability, the question also becomes "how can these communities continue to be viable under a sustainable water use regime". The successful outcome of this process is determined by the combined actions of individual irrigators. Do they have the financial and human capital to make the necessary adjustments? Do they adopt new technologies, management practices and crop varieties to maximize yield so that economic output can be maintained by the use of less water? These questions are relevant for both individual irrigators and irrigation developments or districts. The latter case also includes financial viability and the ability to collect fees, carry out maintenance, etc. (Easter and McCann, Chapter 3). When it comes to the irrigation communities, the issues are much the same. Do they have sufficient social capital to successfully navigate the adjustment process?

Irrigators and their communities, however, do not do this in isolation. Institutional structures have a significant role in facilitating and encouraging this process. Governments can introduce a number of incentives, instruments and policies that will assist the process of adjusting the irrigation industry towards sustainability by (1) encouraging and supporting some irrigators to become viable for long term while encouraging and supporting others to exit the industry and (2) supporting irrigation authorities' ability to maintain sustainable irrigation systems. Some of the policies introduced for these purposes are water pricing, water markets, water user organizations, subsidies for technology and management adoption, structural adjustment assistance, information and education campaigns, and funding research and development. To increase the effectiveness of these policies, the government can put institutions in place which facilitate their adoption (see Chapters 3, 4 and 5). For example,

- For water markets to operate efficiently, the following are needed: (1) well-defined property rights to water so that buyers and sellers know what they are trading in; (2) flexible and secure trading mechanisms to facilitate fast, reliable and secure transactions with minimum transaction costs; (3) properly regulated market intermediaries to facilitate the process; (4) secure water rights registers to provide irrigators with security of ownership and buyers with security of market outcomes and (5) transfer processes, or other regulatory institutions such as water use right, to ensure that trade in water does not have negative environmental impacts so that potential externalities associated with the transfers are internalized. Within some watersheds, special attention needs to be paid to rural-to-urban transfers to make sure that these take place with minimum impact on irrigation (Byrne *et al.*, Chapter 10).

- For water prices to be efficient in contributing towards sustainable irrigation, they need to be set at a level where the necessary infrastructure can be adequately maintained and renewed, and adequate funds are available to properly monitor, enforce and administer water allocations, water scheduling and delivery, fee collection and water trading. They also need to send proper signals to water users. If prices are kept low using subsidies, inefficient and low-value users will continue to irrigate. Full cost recovery prices are more likely to communicate the message to inefficient and low-value users that they need to either become more productive or sell their water to more efficient and higher value users. Water prices could also be structured so that they internalize the externalities associated with water trading out of individual supply systems and thereby help to maintain financial sustainability, as is the case in Australia (Chapters 4, 5 and 11).
- Subsidies have been widely used in the context of irrigation; most large-scale irrigation projects would not exist had it not been for subsidies. In the context of sustainable irrigation, subsidies have been used to encourage the adoption of more efficient irrigation technology. However, great care has to be taken to design such programs so that the efficiencies gained do not result in a net increase in water use with the effect that less water is returned to the river and the environment. There is a growing body of literature illustrating that this has often been the case (Scheierling *et al.*, 2006; Bjornlund *et al.*, 2009).

3 An outline of the book

This book is written within the context of the above-mentioned discussion. Chapters 3–5 deal with the issues of governance and institutions, and, as such, set the backdrop for Chapters 6–15, which deal with individual elements or aspects in more detail.

Chapter 2 (Bjornlund and Bjornlund) provides a historical perspective on sustainable irrigation by discussing some old irrigation cultures and highlighting how some were sustainable while others were not. In some instances, similar irrigation technologies and approaches were sustainable for thousands of years in one location, while they were unsustainable in another, highlighting the location-specific aspect of sustainable irrigation.

Chapter 3 (Easter and McCann) discusses institutional arrangements for sustainable irrigation developments as one of the essential components of sustainable irrigation. It argues the importance of financial sustainability and the need for strong water user organizations to ensure that water charges are paid, allocations are enforced and maintenance work is carried out to keep the systems up to acceptable standards, which limits the environmental impact of irrigation so that the affected water source stays within ecologically sustainable levels as locally defined.

In Chapter 4, de Loe and Bjornlund discuss the governance structures needed to ensure sustainable water management which needs to underpin sustainable irrigation in a contemporary society. They also discuss the inevitable link between sustainable irrigation, water security and food security.

Young and McColl (Chapter 5) then discuss how a robust framework for water allocations can be designed so that it can cope with the change. In the sustainability literature, the need for flexibility and adaptive capacity of institutions has been widely acknowledged. The ability of water resources to meet the need of future generations is dependent on current and future generations' ability to respond to climate change, improved knowledge about environmental needs and changing community values towards water.

To achieve sustainable water management and sustainable irrigation at the watershed level will require a change in human behaviour (Cullen, 1997; GACGC, 1997). The role of the governance and institutional arrangement discussed in Chapters 3, 4 and 5 is to try and facilitate such a change. However, people's behavioural changes in response to such governance and institutional arrangement and instruments will be influenced by the values, norms and beliefs that they hold (Kuehne *et al.*, 2008; Kuehne & Bjornlund, 2007) as will their perceived legitimacy (Contandriopoulos *et al.*, 2004). Kuehne and Bjornlund (Chapter 6 analyse the values, beliefs, norms and attitudes that influence irrigators' behaviour in response to new policies or instruments and discuss their implications in the context of recent offers to buy water from irrigators within the Murray–Darling Basin to provide environmental benefits.

The next three chapters deal with various aspects of water markets. There have been divided opinions on water markets and their ability to facilitate sustainable irrigation in terms of their environmental and social impacts. It has been argued that water trading has negative social and community impacts by favouring larger and more powerful farmers and moving water out of communities depending on it. Others have argued that markets have provided an important mechanism for weaker farmers to stay in business during periods of financial difficulty. In Chapter 7, Shanahan *et al.* analyse water market prices and activities within the Goulburn–Murray Irrigation District in Victoria, Australia to identify whether water markets help to sustain irrigation during periods of drought. They find that markets have significantly increased both big and small irrigators' ability to deal with drought and financially survive at least the first years of a drought period. Markets have helped irrigators with long-term investments to get enough water to keep their businesses going, keep their dairy herds, maintain their vines and fruit trees and fulfil contractual obligations to deliver produce to processors. In this process, these buyers help many smaller and lower value growers to stay on the farm by purchasing their water for far more than the profit they could have made by using it.

In Australia and the United States, increased emphasis has been placed on using the market to buy water to achieve environmental outcomes. In the context of this book, it is considered that acceptable environmental outcomes are ensured at the watershed level as part of the planning process (Chapter 4). In this context, governments can buy water from irrigators to achieve this objective. Individuals and non-government organizations could also be allowed to use the water market as a mechanism to achieve environmental outcomes in addition to the locally defined acceptable outcome. In Chapter 8, Hadjigeorgalis analyses the outcome of water trusts and environmental water transfers in the United States to evaluate their potential impact on sustainable irrigation.

Water transfers out of irrigation to satisfy increased urban and industrial needs have caused some of the major conflicts over water trading in the United States. However, such transfers are likely to play an increased role in water markets as urban centres expand and water scarcity intensifies. It is therefore important to find mechanisms by which such transfers can be implemented with least possible impact on irrigation and food production. Byrne *et al.* (Chapter 9) discuss the possible mechanisms for inter-sectoral transfers and analyse the potential for water options contracts within the Murrumbidgee Valley, Australia.

Water pricing and water markets are often considered to be the key economic instruments in communicating appropriate market signals to water users to encourage them to use water more productively. Also in Chapter 3, Easter and McCann argue the need for appropriate pricing policies to ensure sustainable irrigation systems. In Chapter 10, Young *et al.* pay particular attention to the issue of water pricing in the light of irrigators' ability to pay for irrigation water.

The price of other inputs into irrigation as well as the opportunity to produce more valuable crops also communicates economic market signals to irrigators to use water more wisely. Recently, the price of energy has increased significantly. As energy is a major input in irrigation, the cost of energy has proven a major factor in driving the adoption of more efficient irrigation technology and management practices (Bjornlund *et al.*, 2009). In Chapter 11, Schoengold *et al.* provide an in-depth analysis of the impact of increased energy cost as well as the increase in biofuel production on sustainable irrigation and food security.

The last three chapters discuss sustainable irrigation within the context of three national settings: Alberta, South Africa and India. In many parts of the world, strategies, plans and policies are introduced to resolve problems associated with scarcity, to ensure increased efficiency and productivity of water use so that new demand for water can be met and economic activity can be maintained or increased while more water is set aside to secure healthy ecosystems and environmental outcomes. Given that agriculture uses more than 70% of water extracted for economic use worldwide, irrigation has to play a pivotal role in achieving such objectives. The province of Alberta in Canada introduced its Water for Life Strategy in 2003. In Chapter 12, Nicol *et al.* analyse the strategy's likely success in achieving its efficiency and productivity goals.

Irrigation in South Africa is associated with a unique set of challenges balancing social, environmental and economic interests within the concept of sustainable water management. Nieuwoudt and Backnberger (Chapter 13) discuss how economic instruments and governance structures are being implemented to provide safe drinking water and water for sanitation and subsistence farming to members of previously disadvantaged groups while protecting economic water use and keeping total extraction within sustainable levels.

More than 30% of all water used for irrigation worldwide is groundwater. This has caused significant aquifer depletion in many places such as the United States, China and India, indicating that sustainable irrigation from that resource is not possible at current levels. This has significant implications for sustainable agriculture and food security. In the last chapter, Msangi addresses this issue.

References

Backenberger, G.R. 2005. Water institutional reforms in South Africa. *Water Policy,* 7: 107–123.

Bjornlund, H., Nicol, L. & Klein, K. 2009. The adoption of improved irrigation technology and management practices—a study of two irrigation districts in Alberta, Canada. *Journal of Agricultural Water Management,* 96: 121–131.

Blomquist, W. & Schlager, E. 2005. Political pitfalls of integrated watershed management. *Society and Natural Resources,* 18(2): 101–117.

Brierley, G., Hillman, M. & Fryirs, K. 2006. Knowing your place: an Australasian perspective on catchment-framed approaches to river repair. *Australian Geographer,* 37(2): 131–145.

Contandriopoulos, D., Denis, J., Langley, A. & Valette, A. 2004. Governance structures and political processes in a public system: lessons from Quebec. *Public Administration,* 82(3): 627–655.

Csurgo, B., Kovach, I. & Kucerova, E. 2008. Knowledge, power and sustainability in contemporary rural Europe. *Sociologia Ruralis*, 48(3): 292–312.

Cullen, P. 1997. Issues in the transfer of R&D outcomes to wetland management. In: Williams, W. (Ed.), *Wetlands in a Dry land: Understanding for Management.* Environment Australia, Land and Water Resources Research and Development Corporation, Canberra, Australia.

Elliott, L. 2004. *Global Politics of the Environment (2nd edition).* Palgrave Macmillan, Hampshire.

Flint, R.W. 2004. The sustainable development of water resources. *Water Resources Update,* 127: 41–51.

German Advisory Council on Global Change (GACGC). 1997. *World in Transition: Ways Toward Sustainable Management of Freshwater Resources. Annual Report 1997.* Springer-Verlag, Berlin.

Gleick, P. 1998a. Water in crisis: path to sustainable water use. *Ecological Applications,* 8(3): 571–579.

Gleick, P. 1998b. The human right to water. *Water Policy,* 1: 487–503.

Henriksen, H. & Barlebo, H. 2008. Reflections on the use of Bayesian belief networks for adaptive management. *Journal of Environmental Management,* 88(4): 1025–1036.

Jordan, A. & Jeppeson, T. 2000. EU environmental policy: adapting to the principal of subsidiarity? *European Environment,* 10: 64–74.

Kuehne, G. & Bjornlund, H. 2007. One size does not fit all—recognizing heterogeneity in Australian farmers. *Proceeding from the Conference of the US Commission on Irrigation and Drainage.* Sacramento, California, September.

Kuehne, G., Bjornlund, H. & Cheers, B. 2008. Identifying common traits among Australia irrigators using cluster analysis. *Water Science and Technology,* 58(3): 587–595.

Postel, S. & Richter, B. 2003. *River for Life: Managing Water for People and Nature.* Island Press, Washington DC.

Scheierling, S., Young, R. & Cardon, G. 2006. Public subsidies for water-conserving irrigation investments: hydrologic, agronomic, and economic assessment. *Water Resources Research,* 42: W03428.

The United Nations Conference on Environment and Development (UNCED). 1992. *Agenda 21: Program for Action for Sustainable Development.* UNCED, New York.

United States Department of Agriculture (USDA). 2008. *Sustainable Agriculture.* http://www.usda.gov/oce/sustainable/agriculture.html accessed on 20 August 2008.

Van pelt, M.J.F, Kuyvenhoven, A. & Nijkamp, R. 1992. *Defining and Measuring Sustainability.* Serie Research memoranda. Vrije Universiteit, Amsterdam.

Walmsley, J. 2002. Framework for measuring sustainable development in catchment systems. *Environmental Management,* 29(2): 195–206.

World Commission on Environment and Development (WCED). 1987. *Our Common Future.* Oxford University Press, Oxford, UK.

Chapter 2

Sustainable irrigation: A historical perspective

Vibeke Bjornlund[1] & Henning Bjornlund[1,2]
[1]*Department of Economics, University of Lethbridge, Lethbridge, Alberta, Canada*
[2]*University of South Australia, Adelaide, Australia*

Abstract

Irrigation has been practised for some 6,000 or 7,000 years in various parts of the world. When sedentary farming emerged in areas along or near major rivers in the Fertile Crescent, the Indus Valley and China, irrigation soon followed to sustain food production through dry periods. Many of the areas suitable for irrigation were also prone to flooding and therefore canals and other structures were needed to supply water for irrigation and to provide drainage during floods. Such structures were well beyond individual farmers and more complex societies therefore emerged. The term sustainability was not known to these first irrigators. They experimented, observed, learned, changed and adapted to their physical environment and changing climate. In some instances, the physical conditions were not suitable for intensive irrigation practices. Hence, for a number of reasons such as salinization, siltation, soil erosion and deforestation some hydraulic societies ceased to exist, while in other instances irrigation continued for millennia until irrigation intensified due to increased population pressure, resulting in changing technologies and practices. In other instances, irrigation has continued in the same location until today. This chapter explores these experiences from around the world and try to draw lessons for today's irrigation communities. We find that decentralizing is better than centralizing, diversity is better than uniformity and local self-ingenuity and self-reliance yield superior results. If exogenous technologies or knowledge are applied, it should be in conjunction with, and adapted to, local conditions, knowledge and cultures.

Keywords: Bali; Egypt; India; Mesopotamia; Oman; Sustainable irrigation

1 Introduction

The concept of sustainable development gained prominence in 1987 with the Brundtland Report *Our Common Future* (WCED, 1987) that refers to sustainable development as development that meets the needs of the present without compromising the ability of future generations to meet their own needs. However, the problem of continuing meeting the current generation's need for food and freshwater has been a challenge for mankind since the beginning. Originally, humans lived like hunter-gatherers able to secure adequate food from natural sources. As population grew and also as climate changed in parts of the world, human started to settle down and make a transition to sedentary farming, growing crops and keeping animals (Diamond, 1997). The first examples of sedentary farming are found along some of the world's major rivers in Mesopotamia, or the Fertile Crescent, the Nile, the Indus Valley, in China along the Yangtze and Yellow rivers, and in Mesoamerica. The new farmers soon learned that production could be increased if water was added in the absence of natural precipitation (Hassan, 1997). Under these circumstances, farmers could produce far more food than what they needed for their own subsistence. This made it possible to release large human resources from food production and enabled the emergence of full-time craftpersons and managers who could spend their time and energy on securing food production through the construction of more complex large-scale irrigation and drainage structures. This, in turn, required more complex and organized societies with a division of labour, laws, political hierarchies, etc. The development of larger scale irrigation therefore coincided with the first urban societies. These have been termed hydraulic societies (Wittfoget, 1957). Well known among them are Sumerian, Egypt and the Harappan Civilization in the Indus Valley. The first known written law, the law of Hammurabi from around 1,784 BC, includes 300 sections which dealt with various aspects of irrigation rules about rights and obligations with respect to irrigation water and infrastructure (El-Yussif *et al.*, 1983).

Some of these societies experienced problems with their irrigation practices creating salinity and waterlogging problems, resulting in declining yield and collapsing societies. Others continued for millennia until population pressures or confrontation with exogenous technologies resulting in the adoption of irrigation practices which were unsuitable for the local conditions, and resulted in unsustainable irrigation. Finally, some are still functioning today.

This chapter will discuss some of these early irrigation societies and their changing fortune and draw some conclusions about why some failed while others continued to prosper. Finally, some lessons and implications from these societies will be discussed in the light of today's efforts to secure sustainable irrigation while ensuring food and water security for current and future generations.

2 Examples of early irrigation systems/societies

The following discussions of early experiences with irrigation are not intended to provide a full account of each system; that would be well outside the scope of this chapter. What we will attempt is to explain aspects of each systems from which we believe lessons can be learned for today's sustainability challenge.

2.1 Mesopotamia

Irrigated agriculture in Mesopotamia has been practised along the banks of the Euphrates and Tigris as far back as the fourth or fifth millennium BC (Goldsmith & Hildyard, 1984). Originally, some 10,000 years ago sedentary farming started in areas with sufficient rainfall. However, around 4000 BC local people had moved to dryer areas of Lower Mesopotamia. Initially, they learned to sow their seeds following the spring floods. But in some years, dryness caused the crops to die before harvest. The remedy was simple. They used diversion canals to get the water from the Euphrates and Tigress to their fields and also enabled them to increase production onto areas not covered by the spring floods. They practised irrigated basin agriculture which included alternate-year fallowing during which the land is temporarily abandoned with the results that wild plants take over and dry out the subsoil to a depth of 2 m, thus lowering the water table and preventing the salt from coming to the surface (Jacobsen, 1957–58).

Sometime during the third millennium, there seems to have been a massive increase in irrigation works in the Euphrates Valley, to satisfy the needs for water and food for an expanding urban society such as Ur. The construction of vast canals to supply water to the cities led to seepage which combined with the abandonment of fallowing and over-irrigation led to a rise in the groundwater level. Based on temple records of the time, this seems to have resulted in increased soil salinity (Jacobsen & Adams, 1958).

A particularly devastating impact was experienced following the building of a very big canal, around 2400 BC in southern Iraq by King Entemenak of Girsu. The system consisted of a complex network of canals skilfully planned and regulated. The methods employed appear to mainly involve gravity flow from the Euphrates to irrigate land previously supplied from the Tigris. The canal was so large that it was simply referred to as "the Tigris" (Adams, 1958). It resulted in seepage, flooding and over-irrigation with a consequent rise in the groundwater level – all contributing to increasing soil salinity levels. As a consequence, there was a gradual but marked reduction in the production of salt-sensitive wheat, which was the preferred food crop, and an increased production of the more salt-tolerant, but less attractive barley. Archaeological evidence suggest that equal amounts of wheat and barley was grown in Southern Iraq during the period up to 3500 BC. However, by 2500 BC, wheat accounted for only one sixth of production, by about 2100 BC it accounted for no more than 2%, and by 1700 BC no wheat was grown at all. Not only did they suffer a shift from wheat to barley but soil fertility also declined dramatically. Around 2400 BC, yields were about 2,537 kg per hectare; by 2100 BC it had declined to 1,460 kg, and by 1700 BC to 897 kg per hectare. Inevitably, the large-scale, centralized irrigation systems broke down under the strain of social upheaval (Jacobsen & Adams, 1958).

The Babylonian King Hammurabi (1711–1669 BC) introduced the first set of written laws known as the "Code of Hammurabi". Reflecting the importance of irrigation, the code contained about 300 sections dealing with various aspects of irrigation. One of the laws said:

> if a man is negligent in strengthening the banks of his field and has not maintained his banks and then a breach has occurred in his bank and so he has let the waters carry away (the soil on) the water-land, the man in whose bank the breach has occurred shall replace the corn which he has (caused to be) lost. (El-Yussif *et al.*, 1983)

E. Goldsmith also talks about the modern experience with irrigation under the British colonial system as causing further waterlogging and salinity rendering a third to a half of the land uncultivable. He concludes that

> when (large-scale irrigation) regions are left in the uncontrolled possession of a landlord class, which is either of foreign origin or partner in a precarious alliance with a foreign conqueror, rural investments are in danger of being neglected, because the landlords inevitably go for quick profits and liquid assets. In extreme cases the result is starvation and depopulation. (Goldsmith & Hildyard, 1984)

As a result of all of these factors, it has been argued that out of the total of 48 million dunums of arable land in Iraq only 16 million is cultivable today (Ministry of Irrigation (Iraq) & Selkhozpromexport (USSR), 1975).

2.2 Egypt—The Nile

About the same time as in Mesopotamia, farmers started irrigating along the River Nile utilizing the flash spring floods to inundate their fields in the narrow and low-lying Nile Valley. It was mainly a natural phenomenon but farmers dug flat-bottomed basins along the river and a system of canals to flood the fields from the upstream ends of the basin and drain the water back into the river at the downstream end. The canals were also used to convey the water to fields further away from the river. However, this was limited to the low-lying areas of fertile soil, which could be reached without lifting devices. Inundation irrigation worked sustainable for 5,000 years because of its simplicity, efficiency and cost-effectiveness. The Nile performed three important functions: (1) provided irrigation water, (2) disposed of drainage water and (3) rejuvenated soil fertility by depositing silt from the upstream areas containing important nutrient and minerals. Further, large volumes of water were used only once a year, which leached potential salt deposits out of the soil (Newson, 1992; Population Reference Bureau. 1998).

This system supplied Egypt with ample food and fibre until the 18th century. The population and irrigated area in Egypt prior to 1800 peaked in about 100 AD when Egypt supplied the Roman Empire with vast quantities of grain (Postel, 1999). In the 19th century, Egypt experienced a population explosion and food production for the first time started to fall short of needs. To expand irrigation and food production, a series of dams were built across the Nile north of Cairo followed by a number of other dams culminating when British engineers built the first Aswan Dam in 1904, which was expanded twice by 1934 and completed with the Aswan High Dam constructed from 1964 to 1970.

These developments fundamentally changed irrigation and the flow of the River Nile. Annual inundation and cropping was replaced by continued irrigation and two or three crops a year. The Nile was controlled and did not flood any more. Instead the fields were continuously wet and were never thoroughly leached to flush out the salt from the soil; also the water did not any longer contain nutritional silts as it was trapped behind the dam walls. This resulted in dropping fertility, a built-up of soil salinity and waterlogging (Hunt, 1987).

Already in 1928, water scholars warned about the threat of salinization of the land and that the big dams would hold back the silt that had replenished the fertility of the flood plain for millennia, and cotton harvest started to decline (Carrier, 1928).

2.3 The qanat system

For at least the past 2,500 years underground water mobilization systems, known as qanat systems, have been constructed and used throughout the Arab world. While originating in Iran, the system spread throughout the Middle East, the Arabic Peninsular, North Africa and Afghanistan as well as along the old Silk Route (English, 1968). The discussion in this chapter is based on qanat systems in Oman, where they are locally known as falaj systems.

The system extracts water from the top of aquifers high up in the mountains and brings the water in underground tunnels down to the lower areas and from there often in open channels into villages and onto the small fields, mainly for the irrigation of date palms, but also for other cash or subsistence crops (Al-Marshudi, 2007; Zekri & Al-Marshudi, 2008). This system has several unique features for sustainable irrigation from environmental, social and economic perspectives.

Per definition, the system cannot mine the aquifer as the off-take from the aquifer is located rather high in the aquifer profile; thus, the system only harvests the overflow or the renewable yield from the aquifer. The system requires skill to establish and maintain, but it is entirely gravity–fed, so there is no pumping cost and potential disruption caused by shortage of power or spare parts. This reduces the monetary expenses associated with running the system and provides a secure water supply.

In most systems, only a small proportion of the water rights are privately owned. For example, in one system that one of the authors visited in 2007 only 15% of the water rights are privately owned by the families which originally constructed the system; the rest is owned by the village community by about 10% and the rest by semi-public charity organization. The water which is owned by the village community is sold to the farmers on auctions every week, and the proceeds are used to finance the maintenance and administration of the system.

The majority of the water which is owned by semi-public institutions is sold on annual auctions and the proceeds used to pay schools, police, priest and the poor.

In a sense, this system takes the full consequence of the fact that all wealth is derived from the availability of water and hence the water is sold to farmers to fund the maintenance of the system and the running of the community. This system resolves many of the issues facing irrigation system in developing countries:

1. The problem of fee collection and therefore adequate funding of operation and maintenance, resulting in inefficient systems with reduced productivity as discussed by Easter and McCann in Chapter 3.
2. Equitable distribution of the wealth generated by the water. In this system, this is ensured as most farmers have to buy the water each year with the revenue used to pay communal expenses.
3. Risk sharing associated with reduced supply during periods of scarcity. This is also an integral part of the system as farmers have the right to the flow from the system during a specified period of time. During periods of scarcity, the flow during that period of time simply yields less than during periods of plenty. Hence, the scarcity is shared. The weekly auctions can then be used to redistribute who gets access to the limited flow.
4. Water for domestic use. This is also supplied by the system; as canals enter the village, there are access points where villagers can get access to water for drinking and household purposes; further downstream on the canals are places where households are allowed to wash their clothes after which the water is diverted into the fields. So, all households have access to water for drinking and sanitation before the flow is shared for agricultural purposes.

This system has operated for thousands of years (Al-Marshudi, 2007; Zekri & Al-Marshudi, 2008). In Oman, more than 3,000 such systems were in operation. Today, many have run dry and are not operating anymore. Others are currently under threat. So, why is this otherwise totally sustainable system under threat? In most instances, this is due to the influx of modern technology. Private farmers are sinking tube wells in areas influencing the water level in the aquifers from which the qanat systems are supplied. Once the water level in the aquifer drops below the off-take for the qanat system, the flow of water ceases and the system runs dry, stopping the supply of water for both domestic and farming purposes for the traditional users as well as the flow of revenue supporting the communities.

2.4 Bali—Rice growing

In Bali, intensive rice production has been carried out under irrigation for more than 1,000 years using a community-based water management approach called the subak system (Suarja & Thijssen, 2009). This system is a unique combination of impressive engineering works and complex social structures.

Tunnels, aqueducts and bamboo pipes carry water across deep ravines and through mountains into lakes from where gravity feeds a series of terraced rice fields. These fields have been very productive over centuries producing more than 1 ton of food per acres, making it possible to feed a very dense island population of 2 million people.

The system was, and is, managed by a hierarchy of Hindu priests and water temples. At the top of the system is the high priest, the overall manager of the irrigation system. The high priest is at the main water temple at the Crater Lake near the peak of the system. Below the Crater Lake are a number of major dams, each with a Hindu priest and a water temple, responsible for regulating water flows. The lowest level of the irrigation system consists of smaller weirs, each with a minor water temple to regulate water flows (Suarja & Thijssen, 2009).

The actual irrigation of the fields is managed by 1,500 fully autonomous local farmer organizations (or water user organizations) called 'subaks' to ensure reliable, fair and equitable distribution among its members. The system supports a total of about 90,000 hectares of rice paddy.

A general assembly, of which all 100–200 farmers are members, is the highest authority of each subak. The general assembly elects the board of the subak which is in charge of the daily operation and management of the local system of water distribution among its members. The general assembly facilitates open and democratic discussions among the subak members on issues of general interest in order to generate consensus decisions. Even though the subak system is organized by Hindu priests, all farmers within the subak are members with equal rights irrespective of their faith (Suarja & Thijssen, 2009).

All members of the subak have access to an equal share per hectare of the available water in return for contributing equally in conducting maintenance and other communal work on the subak system. This includes the construction and maintenance of canals, tunnels, aqueducts and dams, and the distribution of water. There are both active and passive members; the active members participate in the maintenance while passive members pay a fee in lieu of work. If the farmer leases the land, then the lease contract will set out who is responsible for the maintenance obligations.

The subak board also coordinates the planting of the rice, which is an essential part of the success of the system as it ensures the best use of the water flow but also an integral part of successful pest control. Planning the planting, harvest, rice burning and the water level after harvest in a coordinated way ensures that the population dynamics of the rice pest is broken. Also, a coordinated effort of breading ducks in the canals and fields helps control weed and pests. This coordination of water use is mutually beneficial to upstream and downstream users, as without it upstream users would suffer from weed and pest and downstream users would have an uncertain water supply. This avoids many of the problems experienced in other systems where downstream users suffer from the actions of upstream users.

This traditional system was left alone for more than 1,000 years, and the Dutch colonial powers did not interfere while they were ruling Indonesia. However, during the Green Revolution of the 1970s the Indonesian government put pressure on the Hindu priests and the subaks to adopt the miracle seed. This included the use of fertilizers and pesticides, three annual crops rather than two, more intensive water use and a disregard for the old communal management system. Initially, production increased but soon production dwindled and cost soared due to water shortage and pest infestation. The farmers lobbied the government to return to the old system. However, they were criticized for being conservative and the new methods continued. In 1983, the National Science Foundation in the United States funded Dr. S Lansing to examine the temple system. Together with a computer expert, he analysed the traditional Bali water management systems and the effect of different cropping systems using 1,000 years of rainfall data kept by the Hindu temples. He found that the traditional system consistently outperformed any other system. Eventually, the Indonesian government encouraged the farmers to return to their old systems (Lansing, 1996).

2.5 India—A diversity of systems

Archaeological evidence suggests that irrigation systems have existed on the Indian Subcontinent for at least 5,000 years. Reflecting the wide variability of rainfall, temperature, altitude, topography and soil types, the Subcontinent has seen a wide variety of early settlements devising their own unique systems of water harvesting adapted to local conditions (Aggarwal & Narain, 1997).

For example, in southern parts of India tens of thousands of village tanks were constructed in the medieval period. This suited the local conditions in the region which were dominated by (1) scanty rainfall, (2) steep slopes with fast run-off and (3) soils with low water retention. There was, therefore, a need to be able to retain and store the water when it was available and apply it when the soil needed it and at a rate at which the soil could absorb it (Aggarwal & Narain, 1997). In other parts of the country, the communities have devised various other means of water harvesting from rivers, aquifers and rainwater run-off to provide water for drinking and agriculture.

There is evidence to suggest that these traditional systems were very efficient and productive. A British survey of 2,000 villages outside present-day Chennai in the province of Tamil Nadu found that between 1762 and 1766, there were villages which produced up to 12 tons of rice per hectare, a level of productivity which today has only been achieved in the most successful areas during the most productive era of the Green Revolution (Hinduism Today, 1997). The same report also found that the availability of food per household in that area was 5 tons compared to today's three-quarter tons. Emphasizing the productivity and efficiency of farming and irrigation in India is the report in the Ecologist magazine by a British agricultural expert. He wrote the report based on a tour to India that he conducted in the 19th century to see how he could best advise Indian farmers on how to improve their agricultural practices. His conclusions reported in the Ecologist magazine stated that he could learn more from the Indian farmers than they could learn from him, as they knew a lot about soil composition and health, pest control and water management, crop breeding and all other aspects of agriculture (Sahtouris, 2009).

The traditional irrigation systems suffered during three periods. During the period when the East India Company ruled India, their primary interest was to extract the maximum level of resources from existing infrastructure and therefore they failed to invest in improvements and neglected the maintenance of infrastructure such as irrigation system. To make matters worse, transport infrastructure such as roads and railways was often constructed on embankments which interrupted the traditional contour-based irrigation canals and tanks. It also cut off large areas from supply of loam-bearing water from rivers, which left the land unproductive and sterile (South Asian History).

During the early days of the British colonial rule, there was some investment in irrigation, such as weirs, dams and barrages, mainly in the Punjab region to produce export crops such as cotton, opium, sugar cane and wheat which generated most revenue to the state (Davies). On the other hand, the British showed no interest in the traditional small-scale peasant irrigation systems of wells, dams, small canals and tanks which had fed the Indian population for thousands of years. Various actions of the British colonial system were detrimental to the continued use of traditional irrigation systems: (1) British tax policies heavily penalized peasants who sank their own wells at their own expense; (2) the general highland tax level during the early colonial period left local communities with no financial ability to maintain their irrigation systems; (3) the taxation system destroyed the social fabric and system which had enabled local communities to construct and maintain irrigation system themselves. Indeed, British rule, in various ways, removed local political chiefs from their obligations to invest in community resources and public institutions such as tank systems; (4) modern irrigation system when constructed were, in many instances, deliberately situated in ways that supplanted other prevailing irrigation structures (D'Souza, 2006) and (5) in some instances, local farmers were actually forbidden to use their traditional system so as to ensure their payment to the new systems (D'Souza, 2006). The refusal of the state to support local irrigation became a smouldering grievance everywhere in interior India (24). This disruption of the community structures was detrimental as traditionally irrigation systems in India have always been treated as communally managed common resources (Hardiman, 1998).

Since independence, the central government of India has shown relatively little interest for traditional small village irrigation systems. The government has been far more interested in promoting the construction of large dams and enhancing the control over the resources by taking over the village tanks. This led to a further breakdown of the community control and management practices that were so vital for the sustainability of the system. A substantial number of tanks acquired by the state have fallen into disrepair because village communities have stopped contributing the voluntary labour for desilting and upkeep of the tanks and the state is not able to perform or pay for the functions earlier executed by the people. By doing this, the state also showed that it is better to construct new large dams rather than revive the efficient network of the village tanks across India (Sengupta, 1991; 1996; Shankari, 1991; Singh, 1994; Von Oppen & Subba Rao, 1980).

This impact was amplified by the fact that the government was committed to the new infrastructure and promoted its use through a set of subsidies which made it more viable for the farmers to use the new systems than maintain the old systems. This has happened in a number of ways: (1) subsidies for electricity supply making it cheap to pump water from dams via canals or from tube wells deep in the ground; (2) subsidies for fertilizers which made farmers less dependent on the use of the tank silt which was used in the past as production-enhancing material; (3) subsidies for water from the major dams which reduced the dependence on tanks for irrigation and (4) subsidies for diesel pumps which also reduced the dependence on water from tanks (Fan & Gulati, 2007).

3 Conclusions

Irrigation has been practised for more than 6,000 years in various parts of the world to feed an ever-growing population. In many places, irrigation was initially sustainable for many years while farming practices were used, which allowed the land to recover between crops and irrigation applications so that salinity build-up was prevented. Once irrigation intensified and continued irrigation and cropping was introduced, the land did not have time to recover and salinity and waterlogging problems soon started to reduce productivity and, in some instances, resulted in the collapse of societies. In Mesopotamia, this started relatively early while in Egypt it did not start until the last century.

In Oman, the qanat system worked both economically and socially and environmentally sustainable up until recent time when the spread of modern tube wells has caused the closure of many systems as water tables are drawn down below the off-take of the qanat system. In Bali, the introduction of the Green Revolution technology soon had disastrous impacts. However, in this instance the government was convinced in time to revert to the sustainable practices of the past.

In India, British colonial rule, in a variety of ways, destroyed many traditional irrigation systems which had successfully fed the local populations for many generations. It destroyed or disrupted no only the physical infrastructure but also the social and community structures that kept these systems in operation. The change in priority from feeding the local population to maximizing export production and the revenue from land taxes was the major reason for this development. Since independence, the Indian government has continued the destruction of the local systems through its emphasis on very large systems and associated subsidy systems favouring the new mega systems at the expense of the local system.

What we can learn from all of this is that we have to respect existing physical and social structures and not replace them with modern technology until we are sure that we fully understand how they operate and how the modern technology is going to interrupt the existing systems and structures. Pursuit of short-term economic gains does not justify causing environmental damage as such harm imposes social and economic costs in terms of higher health bills or diminishing agricultural returns on future generations (Goldsmith, 1998). The most successful systems rely on local knowledge and organization. We conclude that;

- decentralizing is better than centralizing;
- diversity is better than uniformity and
- self-ingenuity and self-reliance yield superior results.

If exogenous technologies or knowledge are applied, it should be in conjunction with, and adapted to, local conditions, knowledge and cultures, drawing on generations of knowledge of, and appreciation of, the local conditions.

References

Adams, R.M. 1958. A synopsis of the historical demography and ecology of the Diyala River basin, central Iraq. In: Woodbury, R.B. (Ed.) *Civilizations in Desert Lands.* Anthrpological Papers No 62, Department of Anthropology, University of Utah, Utah.

Aggarwal, A. & Narain, S. 1997. *Dying Wisdom: Rise, Fall and the Potential of India's Traditional Water Harvesting Systems*, State of India's Environment. A Citizen's Report No. 4, Centre for Science and Environment, New Delhi.

Al-Marshudi, A.S. 2007. The falaj irrigation system and water allocation markets in Northern Oman. *Agricultural Water Management*, 91: 71–77.

Carrier, E.H. 1928. *The Thirsty Earth.* Christophers, London, UK.

D'Souza, R. 2006. *Water in British India: The Making of a 'Colonial Hydrology'.* History Compass 4/4, Blackwell Publishers UK, pp. 621–628.

Davis, M. 2001. Late Victorian Holocausts: EL Niño Famines and the making of the third world, Verso, London.

Diamond, J. 1997. *A Short History of Everybody for the Last 13,000 Years.* W. W. Norton & Company, New York.

El-Yussif, F. *et al.* 1983. Condensed history of water resources developments in Mesopotamia. *Water International*, 8: 19–22.

English, P.W. 1968. The origin and spread of "qanats" in the old World. *Proceedings of the American Philosophical Society*, p. 112.

Fan, S. & Gulati, A. 2007. *Investment, Subsidies, and Pro-Poor Growth in Rural India.* IFPRI Discussion Paper 00716, International Food Policy Research Institute, Washington DC.

Goldsmith, E. 1998. Learning to live with nature: the lessons of traditional irrigation. *The Ecologist*, 28(3): 162–170.

Goldsmith, E. & Hildyard, N. 1984. *The Social and Environmental Effect of Large Dams,* Chapter 25. Wadebridge Ecological Centre, Cornwall, UK.

Hassan, F.A. 1997. The dynamics of a riverine civilization: a geoarchaeological perspective on the Nile Valley, Egypt. *World Archaeology*, 29(1): 51–74.

Hardiman, D. 1998. Well irrigation in Gujarat: systems of use, hierarchies of control. *Economic and Political Weekly*, p.1537.

Hunt, R.C. 1987. Agricultural ecology: the impact of the Aswan High Dam reconsidered. *Culture & Agriculture*, **8(31): 1–6.**

Hinduism Today. 1997. India – Once Plentiful: Records reveal British schemes diminished crops and dismantled a native system of abundance. May 1997. Accessed on 21 August 2009 at http://www.indianscience.org/essays/t_es_crops.shtml.

Jacobsen, T. 1957–58. *Salinity and Irrigation: Agriculture in Antiquity*. Report on Essential Results, Diyala Basin Agricultural Project, p. 85.

Jacobsen, T. & Adams, R.M. 1958. Salt and silt in ancient Mesopotamian agriculture. *Science*, 128(334): 1251–1258.

Lansing J.S. 1996. Simulation modeling of Balinese irrigation. In: Mabry, J. (Ed.) *Canals and Communities: Small-Scale Irrigation Systems*. University of Arizona Press, Tucson, AZ, pp. 139–156.

Ministry of Irrigation (Iraq) and Selkhozpromexport (USSR). 1975. *General Scheme of Water Resources and Land Reclamation*. Ministry of Irrigation, Baghdad.

Newson, M. 1992. *Land, Water, and Development: River Basin Systems and Their Sustainable Management* (Early 1800s population). Routhledge, London, UK.

Population Reference Bureau. 1998. *World Population Data Sheet, Wall Chart* (current population). Washington DC

Postel, S. 1999. *Pillar of Sand: Can the Irrigation Miracle Last?* W.W. Norton & Company (Worldwatch Book), New York.

Sahtouris, E. 2000. Earthdance: Living Systems in Evolution, Chapter 20: Sustainable Society. iUniverse. Accessed 21 November 2000, ISBN: 0-595-13067-4 at: http://www.ratical.org/LifeWeb/Erthdnce/chapter20.html

Sengupta, N. 1991. *Managing Common Property: Irrigation in India and the Philippines*. Sage Publications, New Delhi.

Sengupta, N. 1996. The indigenous irrigation organisation in South Bihar. In: Barah, B.C. (Ed.) *Traditional Water Harvesting Systems——An Ecological Economic Survey*. New Age International Publishers, New Delhi.

Shankari, U. 1991. Major problems with minor irrigation, *Indian Sociology*, 25(1), p. 85–111.

Singh, K. 1994. *Managing Common Pool Resources: Principles and Case Studies*. Oxford University Press, Delhi.

South Asian History, *History or the Indian Subcontinent: British rule and the legacy of colonization*. Accessed on 21 August 2009 at: http://india_resource.tripod.com/colonial.html.

Suarja, I.G. & Thijssen, R. 2003. Traditional water management in Bali. *Leisa Magazine*, pp. 25–26. Accessed on 21 August 2009 at www.metafro.be/leisa/2003/193-25_26.pdf.

Wittfoget, K.A., *Oriental Despotism: A Comparative Study of Total Power*. Yale University Press, New Haven, p. 239.

World Commission on Environment and Development (WCED). 1987. *Our Common Future*. Oxford University Press, Oxford, UK.

Von Oppen, M. & Subba Rao, K.V. 1980. *Tank Irrigation in Semi-Arid Tropical India*. Mimeo, ICRISAT, Patancheru, AP, India.

Zekri, S. & Al-Marshudi, A.S. 2008. A millenarian water rights system and water markets in Oman. Water International, 33(3): 350–360.

Chapter 3

Sustainable water projects: The task of economic instruments and supporting institutions

K.W. Easter[1] & L.M.J. McCann[2]
[1]*Department of Applied Economics, University of Minnesota, MN, USA*
[2]*Department of Agricultural Economics, University of Missouri, MO, USA*

Abstract

In this chapter, a number of critical institutions are identified that improve the financial sustainability of irrigation and water projects over time. The framework for analysis is based on Williamson's four levels of institutions (Williamson, O.E. 2000. The new institutional economics: taking stock, looking ahead. *Journal of Economic Literature*, 38(September): 595–613). The four levels are used to highlight the problems that arise in designing institutions and implementing institutional change. Examples are provided of institutional reforms and changes that have helped different countries raise both cost recovery rates and rates of collection. A key objective in designing water instruments is to provide water users and managers assurance regarding the actions of others within the irrigation project. Without the appropriate institutional setting, it will be difficult to effectively use different economic instruments, such as water prices, taxes, quotas, or markets, to improve the financial sustainability of water projects.

Keywords: Cost recovery; Economic instruments; Financial sustainability; Institutional arrangements; Water pricing

1 Introduction

One of the keys to sustainable water systems, particularly irrigation, is the finances to support and maintain water delivery and its associated drainage or disposal system. Historically, the lack of adequate finances has resulted in inadequate system operation and maintenance (O&M) and caused many water systems to be built with inadequate delivery and control structures and, in many cases, no facilities for drainage and wastewater disposal. The end result has been projects that decline rapidly in their ability to provide adequate and timely water service. After a number of years, many irrigation projects face declining irrigated acreage as waterlogging and salinity problems force agricultural land out of production (Ward, 2002, pp. 107–108). Thus, once it has been determined that it is appropriate to build, and benefits are estimated to exceed costs of an irrigation or water project of a given size and design, we need to determine how to appropriately fund the project over time. The key questions we will try to address in this chapter are what share of project costs can reasonably be paid by different groups of water users and how this share can be effectively collected on a sustainable basis.

Setting up a system that will provide sustainable funds for water projects, including the necessary drainage and water disposal infrastructure, will involve establishing effective institutional arrangements to support efforts to collect water charges from users. Institutions can be thought of as the "rules of the game" while the "players or groups of players" are the organizations, firms, and individuals using, operating, and managing the systems (North, 1990). Institutions are important in structuring incentives as well as providing order and predictability, particularly regarding the actions of others. Livingston and Garrido (2004, p. 25) argue that institutions are important for effective water management. "Institutional arrangements are critical in creating incentives because they (1) define who has access to water resources, (2) establish the range of (legal) options open to legitimate water users, and (3) determine who can claim income from water use and who will bear the costs of water use." Thus, effective institutional arrangements will need to be in place to have sustainable finances and to sustain water systems in general.

The remainder of the chapter will start with a brief description of the institutional framework used in the analysis. This is followed by a section dealing with fee collection and the determination of how much users should pay. Next, is a discussion of water pricing mechanisms, followed by examples of projects where new institutional arrangements have helped improve cost recovery from users, resulting in improved sustainability of irrigation systems. This leads to a section that suggests how the institutions can be combined to provide a stable source of funding for O&M and a more sustainable irrigation system. The final section provides a brief conclusion.

2 Institutional setting

A good way to think about institutions and how they influence outcomes is to use Williamson's (2000) four levels of nested institutions (Figure 1). They include, first, the informal institutions such as social norms, customs, and religion, which change very slowly. These norms and customs constrain or enable what can be done at the other three levels. For example, strongly held customs or values regarding open access to water and free public services may have a large impact on who gets water and how much, if any, they pay. It may prevent or make it difficult to establish private property rights for water or its use and to introduce water markets. Also, values or beliefs concerning the environmental services provided by water will strongly influence an area's efforts to reduce water pollution.

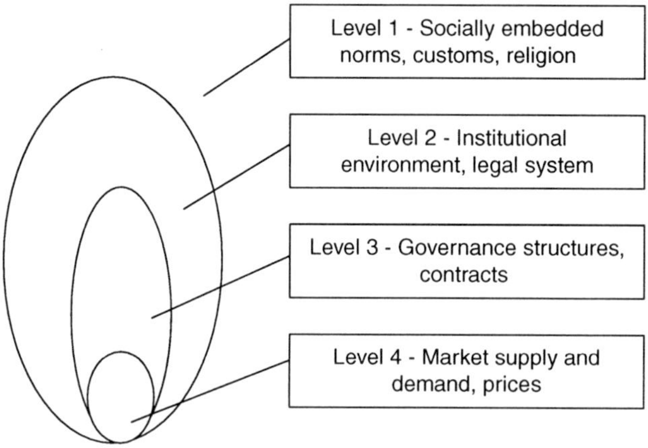

Figure 1: Levels of institutional analysis. Adapted from Williamson, O.E. 2000. The new institutional economics: taking stock, looking ahead. *Journal of Economic Literature*, 38(September): 595–613.

The second level of institutions is the formal rules of the game or the policies that guide water use and allocation. To make changes at this level will usually take several years to over a decade. For example, if you want to establish a water market, one of the needed changes is to establish and allocate water rights or use rights to individual water users. As noted above, such changes in property rights, laws, or policy can be difficult to make. Existing systems are usually changed only after a number of years of hard negotiation and bargaining or after a significant change in a country's economic policy that favors markets as happened in Chile in the 1980s (Hearne & Easter, 1995). The content of water policies and water laws is addressed at this level, including the question of whether water can be sold separately from land and, if it can, what end uses it can be sold for outside of agriculture. In terms of financing water systems, it is this level where water policies are crafted that specify who pays for water projects and their operation, as well as any rules regarding the discharge of drainage water. These latter rules can be critical for the sustainability of irrigation systems facing potential waterlogging and/or salinity problems.

Level three focuses on governance structures for transactions. This is the level where decisions are made concerning the mechanism for allocating water, for example, hierarchy versus markets or contracts. In other words, do water managers, every year, decide who gets water or is this determined ahead of time by water rights or contracts? There will also be concerns about mechanisms for enforcing water allocations and resolving conflicts that are likely to arise over time. The complexity of the governance structures will increase as water scarcity and its value rise and conflicts accelerate. Yet the structures will be constrained to a significant degree by past institutions or what economists call path-dependency. In Valencia, Spain where water is scarce the water court meets every Thursday on the church steps to resolve any water disputes. The past experiences Valencia has had in providing irrigation under scarce water conditions clearly influenced its choice of governance structure. A good experience in providing such collective good makes governance easier. In contrast, a bad experience with cooperatives or other user-run organizations may make user-based governance structures very difficult to establish. Birner and Wittner (2004) argue that the most efficient forms of governance and their structure and complexity will depend on water scarcity and other characteristics of the resource, including the water infrastructure put in place and the social and political characteristics of a country.

The fourth or final level of institutions falls in the domain of neoclassical economics where governance is ignored and the emphasis is on the firm as a production unit. The first three levels of institutions are generally assumed to be fixed and treated as exogenous constraints. At this stage, questions are raised regarding the firm's or water user's ability to pay for water and how water charges or fees may change water use or the adoption of new water-saving technology. Level four institutions are very important in determining the actual level of water charges paid by different types of water users and the services provided to them by the water system. (See Easter & McCann, 2009 for a more detailed discussion of water institutions.)

3 Financial failures in public irrigation

Traditionally, both developed and developing countries have found it difficult to establish a sustainable source of funding for operating and maintaining their irrigation systems and, more generally, their water projects. In his 1995 study, Jones (1995) illustrates how cost recovery and charges for irrigation water have been a problem for decades. In many countries, less than 20% of the cost of irrigation projects has been recovered from water users (Easter & Liu, 2007). This is the result of poorly designed collection rates and practices combined with relatively low water fee levels. The end result has been a large public subsidy for water users, particularly irrigators.

There are many reasons for this poor record of cost recovery in public irrigation projects. Although the reasons vary somewhat among countries and individual projects, Easter and Liu (2007) list some of the most important ones. They are listed below in no particular order of importance and not all are likely to be a problem in each financially strapped water project. "(1) no link between fees collected and funds allocated to a given water project, (2) lack of water user participation in planning and management of projects, (3) poor communication and lack of transparency between water users and management, (4) poor delivery of water services (timing, duration, and quantity are inadequate), (5) no penalties for managers and staff who provide poor service, (6) no penalties for water users who do not pay water charges, (7) low priority given to fee collection, efficient water use, or system O&M, (8) inappropriate infrastructure design and technology to effectively manage water system, and (9) corruption among officials and those collecting water charges." Easter and Liu (2007) go on to make it clear that the basic underlying causes for the poor cost recovery stem from "the collective good nature of water projects, combined with open access to water resources, the principal-agent problems, and rent-seeking activities of water officials. It also can be thought of as an assurance problem: assurance for managers concerning what water users will do and assurance for water users concerning what water managers and their staff will actually do as opposed to what they say they will or can do given the existing project design and technology (Easter, 1993)."

Another part of the problem stems from the fact that many times no one thinks about sustainable finances early enough in a project's development. During the planning stage, and before the project is built, decisions need to be made that provide an effective means for financing O&M. As part of this financial planning, it should be determined what share of the costs should be paid by water users and how the cost should be allocated among the various users, for example, irrigators, hydropower users, domestic water users, commercial and industrial water users, and those protected from floods. The allocation of costs is an important issue because many water projects are multipurpose, particularly those in Asia where 90% of the dams for irrigation are multipurpose. There is also a good argument to be made that some of the water project costs should be allocated to consumers who benefit from the lower food costs due to irrigation, particularly in developing countries. In fact, if commodity markets are localized, the increased crop production, due to irrigation, may mean farmers receive only modest increases in net returns because of the drop in commodity prices caused by increased production in the irrigated area. Yet many times decisions have been made to allocate farmers most of the project costs to be recovered, since they receive much of the water. A better allocation of project costs might be based on who receives direct project benefits or who gets direct and indirect project benefits combined. Easter and Liu (2007) show that it makes a significant difference. For example, if the costs to be recovered are allocated based on water delivered in the Sriram Sagar Project in India, farmers must pay 95% of the costs (Table 1). If the costs are allocated based on direct benefits, farmers only pay 88% of the costs (Easter, 2003). The percentage of cost allocated to farmers would drop even more if the allocation was based on direct and indirect benefits. The cost allocated to farmers would probably drop to something less than 75%. In contrast, two other multipurpose projects in Andhra Pradesh, India would still have over 90% of their costs allocated to irrigation even when costs are allocated based on direct benefits (Table 1).

Once a reasonable allocation of costs has been decided on, the next question is, what economic instruments can be used to effectively collect the water fees necessary to cover the allocated costs? An equally important question that needs to be addressed at the planning stage is, what are the critical institutional arrangements that need to be in place, which will help make the fee collection effective? This is important because the design of the project infrastructure may well place limits on the institutional options that are feasible as will existing institutional arrangements. In many cases, work has to start on implementing new institutional arrangements before project construction starts. This is particularly

Table 1: Alternative cost allocation for three water projects in Andhra Pradesh, India.

Cost allocation for three consumptive uses based on water delivery			
Three water projects	*Domestic water supply (%)*	*Industrial (%)*	*Irrigation (%)*
Nagarjunasagar	2	0	98
Tungabhadra	1	4	95
Sriram Sagar	2	3	95
Cost allocation among three projects based on direct benefits			
	Three water projects		
Purpose or use	*Sriram Sagar (%)*	*Nagarjunasagar (%)*	*Tungabhadra (%)*
Irrigation	88.1	94.3	91.3
Hydropower	3.0	4.0	4.2
Domestic	3.0	1.6	2.1
Industry	4.3	0.1	2.3
Fisheries	1.6	0.1	0.1

Source: World Bank, 2003b. *The Incidence of Canal Irrigation Subsides in India,* Annex, South Asia Rural Development Unit, The World Bank, Washington, D.C., p. 57.

true if water rights or new governance arrangements are to be introduced to help allocate water and collect fees.

4 Economic instruments

The approach one selects for charging water users will depend a lot on what institutions already exist and the size of the project in terms of both hectares irrigated and the numbers of farmers actually served. For large projects in developing countries serving a large number of farmers, but with limited farmer participation, *area-irrigated*-based fees have generally been used. The problem is that there is no relationship between what farmers pay and the amount of water they receive. This also means that water charges fail to create incentives for farmers to use less water. A better alternative may be to vary the charge by type of crop grown with higher per hectare water charges for crops such as rice and sugarcane that use more water. In some cases, water charges might be varied based on the irrigation technology used. If farmers adopt improved irrigation technology such as sprinkler irrigation, which distributes water more uniformly across the field than does flood irrigation and uses less water per hectare, they would be charged a lower water fee per hectare.

For smaller projects, particularly in more developed countries that face high water demands relative to supply, we need to be moving toward *volumetric-based* water charges. With recent improvements in technology, arguments that water use or delivery is hard to measure are no longer very convincing. The big issue may be the infrastructure cost, but as Easter and Liu (2007) discussed, even this may not be a true constraint today in areas facing severe water scarcity, given the new technology available (see Wang & Lu, 1999). Volumetric-based water charges have two clear advantages. First, farmers are charged based only on the quantity they receive. Second, it gives farmers an incentive to not overirrigate and to conserve water, since if they use less, they pay less. This, of course, will require not only water measurement and appropriate infrastructure but also effective staff to manage water deliveries.

Block water charges or prices can also be used with volumetric-based fees as a means to provide a minimum amount of water at low prices while charging much higher prices for levels of use exceeding a set volume of water use per hectare (see Figure 2). This type of charge has several benefits. First, it gives water managers at least three instruments to change cost recovery: the levels of the first and second block charges and the quantity at which the second block charge or price starts (e.g., 3,000ℓm^{3}ℓha^{-1} versus 4,000ℓm^{3}ℓha^{-1}). Second, the last block charge can be set at the marginal cost (MC) of new water supplies. The charge based on MC will encourage farmers to conserve water and use it efficiently. Finally, the higher charges will have less of an adverse impact on farmers' income since they are only charged higher rates for the units of water used beyond the first block. Of course, in many surface water irrigation systems impacts on farmers' income may not be a critical issue since water fees usually do not represent more than 2–7% of farmers' net income (Cornish & Perry, 2003).

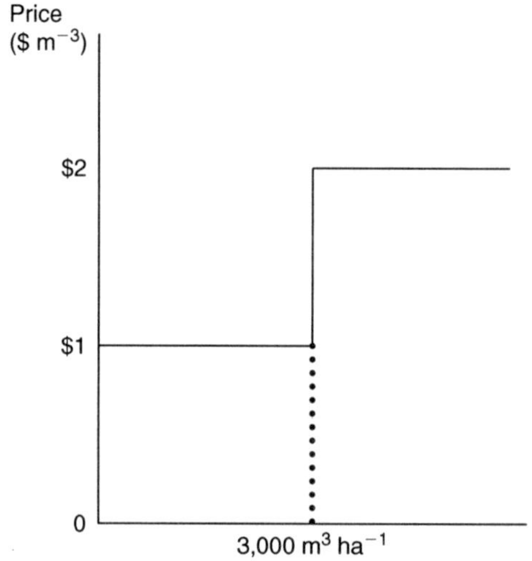

Figure 2: Pricing schedule for irrigation water.

A *two-part charge* can also be used, combining volumetric charges with a fixed charge. Such a system of charges may be necessary if water availability varies a lot from season to season and/or from year to year. The fixed charge allows water managers to obtain a basic amount of funding even in years of low water supplies when many farmers do not receive much water. If water charges were only based on volumetric charges, then in the years of low supplies the volumetric charges might have to be very high to cover project costs. The two-part charge for water also recognizes that there is a large fixed component in O&M that does not depend on the amount of water delivered. These costs need to be paid even if little or no water is delivered.

A final option is to introduce a *water market*. This could be in addition to water charges for O&M. If farmers are allowed to trade water, this will improve allocation efficiency by increasing water's value to farmers and shifting water to its highest valued uses. In turn, this would increase the user's ability to pay since water would be used to produce higher valued crops and/or provide income directly through water sales. As discussed above, the introduction of markets would require a number of key institutions, including a water law that allows the sale of water independent of land and specifies how disputes over water rights will be resolved. In addition, a court or council should be established to resolve disputes over water rights and third-party effects.

5 Institutions to improve cost recovery

One good measure of the success of a water system in sustaining its financial base is to look at the collection rate from its water users. In many systems, only 10–60% of the users pay their fees (Easter & Liu, 2005). Table 2 lists eight irrigation projects that have been quite successful in obtaining high rates of fee collection, with half reaching 100%. In addition, with their improved system of water charges, four systems reported significant water savings (Table 3). Using the projects listed in Tables 2 and 3 as examples of successes in sustaining finances, what institutions were important in their success? There are several key institutional arrangements that have helped these systems move closer to financial sustainability. Institutions do this by helping to correctly align incentives. First, a country needs to be able to legally establish water management entities that are, primarily, financially autonomous from government. This helps create an important set of incentives for water management that will foster improved collection rates. Second, a law is needed that allows users to organize legally into water user associations (WUAs) and participate in management decisions regarding the water allocation. If possible, WUAs should have some authority to allocate water among users and authority to tax them. Through active participation, water users will establish a closer working relationship with management that will help increase system transparency and accountability, which, in turn, will improve users' willingness to pay their water fees. To be effective, WUAs have to be given some authority and important management functions so that they

Table 2: Factors influencing fee collection rate.

Cases	Financial autonomy	Incentives to collect	Incentives to pay		User participation	System transparency	Collection rate (%)
			Penalty for nonpayment	Improved irrigation service			
Awati, China	Yes	Yes	N.A.	N.A.	Yes	N.A.	98
Bayi Irrigation District, China	Yes	Yes	Yes	Yes	Yes	Yes	100
Nanyao Irrigation District, China	Yes	Yes	N.A.	N.A.	Yes	N.A.	95
Shangdong, China	N.A.	N.A.	Yes	Yes	N.A.	Yes	100
Gujarat, India	Yes	Yes	Yes	Yes	Yes	N.A.	100
Haryana, India	Partly	N.A.		N.A.	Yes	N.A.	85–85
Mexico	Yes	N.A.	Yes	Yes	Yes	N.A.	90
Alto Rio Lerma, Mexico	Yes	Yes	N.A.	Yes	Yes	Yes	100

N.A. Not available; 1ℓmu = 0.067ℓha.
Source: Easter, K.W. & Liu, Y. 2007. Who pays for irrigation: cost recovery and water pricing? *Water Policy*, 9: 285–303.

Table 3: Factors influencing water use efficiency.

Cases	Increase per unit price	Switch to volumetric metering	Pricing structure	Water-saving technology availability	Assurance of water delivery	Education		Annual saving
						Public awareness	Technical assistance	
Awati, China	Yes	Yes	Increasing block	N.A.	N.A.	Yes	N.A.	$50 \ell m^3 \ell mu^{-1}$
Shangdon, China	N.A.	Yes	Volumetric	N.A.	Yes	N.A.	N.A.	$5\ B\ell m^3$
Yangtze basin, China	N.A.	Yes	Volumetric	N.A.	Yes	Yes	N.A.	$1.18\ M\ell m^3$ in WUA
Katepurna, India	N.A.	Yes	Volumetric	Yes	Yes	Yes	Yes	$7.71\ M\ell m^3$

Source: Easter, K.W. & Liu, Y. 2007. Who pays for irrigation: cost recovery and water pricing? *Water Policy,* 9: 285–303.

can, and will be, active participants. WUAs need to feel a "sense of ownership" in the water projects (Ward, 2002, p. 102). Too many times, WUAs have been constituted and given only limited responsibilities, such as contributing "free" labor, but no authority or "ownership." This is why a number of WUAs have not improved the sustainability of their irrigation systems (Zekri & Easter, 2007).

However, if WUAs are given adequate authority over system management, labor contributions by members can be used as a partial substitute for higher water charges and help reduce financial constraints. This has been done with limited success in irrigation systems serving small-scale farmers with limited cash income. Paying a share of their water fees with labor can be particularly effective if system maintenance can be done during periods when farm families are likely to have surplus labor. If done right, WUAs can effectively reduce their financial resource requirements through members actively participating in system O&M.

Why is it that establishing financial autonomy and active user participation improves the sustainability of irrigation projects in terms of both their operation and finances? The financial autonomy changes several important incentives. First, the fees collected from users go to the project and do not go back to the state or federal treasury where they would be commingled with other tax returns. In other words, if you do not pay your water fees, it will have an impact on "your" project's ability to deliver water. That is not the case in many countries such as India where the revenue arm of the government collects or tries to collect the water charges, which then go to the federal or state treasury. In such cases, irrigators do not see any relationship between what they pay and the service they receive from their irrigation system. Over time, without enough funding for resources to operate and maintain a project, it will deteriorate rapidly and the sustainability of the irrigation project will be at risk.

The Yangtze Basin Water Resource (Yangtze) Project in China is a good example of an effective water management entity that is financially autonomous. It also requires direct involvement of WUAs in water management decisions. This has helped increase crop yields and saved significant amounts of water (Table 3). The financial autonomy created a second important incentive: to save water. The Yangtze basin authority invested in upgrading its water delivery infrastructure so that it could reduce water losses and save water. This gave the basin authority more water to sell, which increased their cost recovery (Lin, 2003).

Financial autonomy means that subsidies from government are not available and they must rely on fees collected from users to cover O&M costs. This is another important incentive. Now the water management entity will want to create conditions that result in good collection rates from water users. As a result, management will employ several different strategies. One is to use strictly enforced penalties for those who default on payment. For example, in the Bayi Irrigation District, irrigation water was denied to those who defaulted (Johnson *et al.*, 1996). A second strategy is for the management entity to give awards or penalties to encourage staff to achieve higher rates of collections. In Awati, China, they made staff salaries completely dependent on water charges and collection rates reached 98% (Awati County Government, 2002). The Bayi Irrigation District achieved 100% collection rates by combining user penalties with staff rewards for turning in collected fees by a deadline and fines for any late payments (Johnson *et al.*, 1996).

In terms of user participation, it is beneficial to get them involved early in the process of project design and building. This is particularly true for project rehabilitation where farmers are expected to repay the costs or contribute labor. This makes the decision-making process more transparent and increases users' willingness to pay for improvements. The Laur Project in the Philippines provides one example where the WUAs were able to review the rehabilitation proposal before it was implemented. Coward (1980) found that this improved the project design and the users' willingness to pay.

Finally, a water management entity that is financially autonomous has an incentive to provide users a good service. This will not only give users more reason to pay their fees but also increase their ability to pay. Better irrigation service should result in increased yields as it did in the Yangtze and Katepurna Projects and even make it possible to grow higher valued crops. Both should increase farmers' incomes and ability to pay for the water.

Another more dramatic way to increase user participation and create management incentives has been irrigation management transfer (IMT) to the users. This strategy has had mixed results partly because of what management was actually transferred and the condition of the infrastructure at the time of the transfer. Several of the transfers have gone well (Kloezen *et al.*, 1997) while others have not. As Zekri and Easter (2007, p. 573) found, IMT tended to be "successful where farmers had their water rights established, farms were medium and large scale with good access to markets and the government had a strong willingness to empower users." In contrast, programs that emphasized only farmer participation with no decision-making authority over water allocation were not very successful. In other words, farmers need to perceive some clear benefits from participation and taking over the extra work and responsibility. Again the incentives need to be aligned by having the benefits to farmers exceed their costs.

6 Supporting institutions

Too many times the focus is only on the technical or engineering side of irrigation projects. Of course, the engineering aspects of a project are important and, in a technical sense, determine the type of irrigation that is possible. What have been left out are the institutional arrangements that provide the restrictions and incentives to encourage appropriate human actions. These institutions will determine who actually has access to the irrigation, who can claim income from the water, and who will bear the cost of water use. As Livingston and Garrido (2004) point out, the allocation and allotment of water will depend on the institutional arrangements that establish the rules for water use and the payment for this use.

Some of the institutions that are critical for sustainable financing of irrigation project have been discussed above. These include the legal institutions that allow financial autonomous water management entities to be established and effective

WUAs to be created. It is also becoming clearer that water users need some types of water rights that can give them assurance regarding when and how much water they will receive during the crop season. These might be outright water rights that can be traded either permanently or temporarily, or water use rights that have a set life span, for example, 30–99 years. Another option that establishes the right incentives is a water contract, between farmers and the water management entity, that is enforceable and clearly specifies the amount and timing of water to be delivered to farmers. This was one of the keys to the success of the improved irrigation project in Katepurna, India where they have been saving 7.7 million m^3 of water annually and expanded the area irrigated by 80% (Table 3). Overirrigation in the wet season was greatly reduced since the farmers no longer had to store irrigation water in their soil during the wet season so that it would be available to grow crops in the dry season. They now have a contract assuring them water for the dry season (Belsare, 2001).

As part of any institutional arrangement that establishes farmers' rights to water, it is important to have a clear mechanism for allocating the rights. For existing systems, this is not as difficult a question as it is for new systems. In Chile when they established water rights in 1981, the consumptive use rights were allocated based on past water use. This turned out to be fairly equitable since most irrigated land holdings were relatively small, 50 ha or less, and widely distributed among farmers in the irrigated areas (Hearne & Easter, 1995). It is a more difficult question for new irrigation projects. If the area to be irrigated is already farmed, then past land ownership usually has to play a significant role in the allocation of water rights. Yet some, or even all, of the water rights could be auctioned off, particularly in developed countries. Another option is to limit the water any one person can receive or the area they can irrigate. For example, the US Bureau of Reclamation, which has, in the past, funded irrigation projects in the US West, has tried to limit the acres one person can irrigate from a reclamation project to 160 acres or 65 ha. This restriction was never very effective particularly in California even after the 160 acres acreage limit was increased sixfold in the 1980s (Wahl, 1989, Chapter 4).

You also need an effective local system for enforcing water use rules and water rights. In more developed countries, this might be the court system, a water agency, a WUA, or some combination of the three. For example, in Chile, WUAs play a major role in the operations of local systems. In less developed countries, such as China, the village leadership is likely to play a much larger management role. In some cases, the village leadership makes it difficult to establish other methods of managing water by essentially running any WUA. This is a case where existing organizational and institutional arrangements limit what you can do in managing the irrigation project and how you can do it.

Another important set of institutional arrangements that need to be in place to improve fee collection is mechanisms to make the process of setting water charges more transparent. If the charges are based on the cost of O&M, as they are in many cases, then users need to know how the costs are calculated and what costs are to be included. This knowledge can help reduce the fear among some water users that the fees will just be used to enrich the water managers and their families. You do not want water users thinking that an increase in water charges will be used to buy new jewelry for the water manager's wife.

Finally, institutional arrangements are needed that foster the establishment of village councils or boards that can review the record of farmers who are not paying their water fees. This review can serve two important purposes. First, the council can determine if there are circumstances that make it difficult for some farmers to pay their fees such as crop failure. In such cases, the council can then decide to forgive some or all of the farmer's delinquent fees. Second, they can use the review as a time to make public in the villages who is not paying their water fees. Finally, the council could also be responsible for addressing any farmer's complaints raised about management of the WUA.

7 Conclusion

This chapter emphasizes the importance of institutional arrangements in successful irrigation and focuses on those institutions that are critical to a water project's financial sustainability. Williamson's (2000) four levels of institutions are used to show how institutional arrangements at these different levels can guide water use and facilitate the process of determining who should pay for the water. A well-constructed irrigation system only assures that water can be delivered, not that water will be delivered and used effectively or that it will be financially sustainable.

Based on past projects that have been successful in maintaining high rates of fee collection from their users, several institutional arrangements appear to be critical. One is for the management or operating entity to be financially autonomous from government. This creates a set of incentives for management that focus on good service, accountability, and the importance of fee collection. If the farmers are going to pay for the irrigation, they want some assurance that the service will be timely and dependable. One set of institutional arrangements that increase these assurances for farmers are water rights or water use rights. There is also a possibility that a similar assurance can be achieved through establishing contracts between management and water users regarding water delivery and payment schedules. The trick is to make these contracts transparent and binding on all parties.

Another set of institutional arrangements that will be important in maintaining high levels of fee collection are those that improve communication between management and water users. Here, WUAs that are given authority along with responsibility for management can play a key role. They can also be important in making the process of determining and setting water charges more transparent and in helping establish mechanisms to deal with farmers who default.

Finally, financial sustainability of a water system is critical for the overall sustainability of the system. Adequate finances will allow the system to be built in a timely manner with adequate control structures and drainage. Financial sustainability will provide the funding necessary for effective system management and maintenance over time. Third, it will enable management to employ staff that enforces the rules for water use and allocation. This will give water users assurance that the rules will be followed by the other users. Finally, a strong financial position will allow a water management entity to adequately compensate its staff so that the incentives to seek bribes are significantly reduced.

References

Awati County Government. 2002. *Report of Water Management Institutional Reform in Fengshou Irrigation District*. Awati County, Xinjiang, China, 2002, http://www.xjawt.gov.cn/xianqing/jinji/fsgq.htm (in Chinese).

Belsare, E.S. 2001. *Participatory Irrigation Management in Katepurna Irrigation Project: A Success Story*. ICID Watsave Young Professional Award Winning Paper, New Delhi, India, http://www.icid.org/belsare_2001.pdf.

Birner, R. & Wittner H. 2004. On the 'efficient boundaries of the state': the contribution of transaction-costs economics to the analysis of decentralization and devolution in natural resource management. *Environment and Planning C: Government and Policy*, 22(5): 667–685.

Cornish, G.A. & Perry, C.J. 2003. *Water Charging in Irrigated Agriculture: Lessons from the Field*. Report OD 150, HR Wallingford Ltd, Wallingford, UK.

Coward, E.W. 1980. *Irrigation and Agricultural Development in Asia*. Cornell University Press, Ithaca, New York.

Easter, K.W. & Liu, Y. 2007. Who pays for irrigation: cost recovery and water pricing? *Water Policy*, 9: 285–303.

Easter, K.W. 1993. Economic failure plagues developing countries' public irrigation: an assurance problem. *Water Resources Research*, 29(7): 1913–1922.

Easter, K.W. 2003. *Cost Allocation and the Distribution of Costs for Irrigation in Andhra Pradesh*, India. Unpublished paper. Department of Applied Economics, University of Minnesota, MN.

Easter, K.W. & Liu, Y. 2005. *Cost Recovery and Water Pricing for Irrigation and Drainage Projects*. Agriculture and Rural Development Discussion Paper 20, World Bank, Washington DC.

Easter, K.W. & Liu, Y. 2007. Who pays for irrigation: cost recovery and water pricing? *Water Policy*, 9: 285–303.

Easter, K.W. & McCann, L.M.J. 2009. Nested institutions and the need to improve international water institutions. *Water Policy*.

Hearne, R.R. & Easter, K.W. 1995. *Water allocation and water markets: an analysis of gains-from-trade in Chile*. World Bank Technical Paper 315, Washington DC.

Jones, W.I. 1995. *The World Bank and Irrigation*. World Bank Operations Evaluation Study, Washington DC.

Johnson III, S. H., Vermillion, D., Svendsen, M., Wang, X., Zhang, X. & Mao, X. 1996. *Management Reform and Performance Changes in Two Irrigation Districts in the North China Plain*. The Short Report Series on Locally Managed Irrigation, Irrigation Water Management Institute, Colombo, Sri Lanka.

Kloezen, W.H., Garces-Restrepo, C. & Johnson III, S. H. 1997. *Impact Assessment of Irrigation Management Transfer in the Alto Rio Lerma Irrigation District, Mexico*. Research Report 15, International Irrigation Management Institute, Colombo, Sri Lanka.

Lin, Z. 2003. *Participatory Management by Farmers – Local Incentives for Self-Financing Irrigation and Drainage Districts in China.* Environment and Social Development East Asia and Pacific Region discussion paper, World Bank, Washington DC.

Livingston, M.L. & Garrido, A. 2004. Entering the policy debate: the economic evaluation of groundwater policy in flux. *Water Resources Research*, 40: W12S02, doi: 10.1029/2003WR002737.

North, D.C. 1990. *Institutions, Institutional Change and Economic Performance.* Cambridge University Press, New York.

Wahl, R.W. 1989. *Markets for Federal Water (Subsidies, Property Rights and the Bureau of Reclamation).* Resources for the Future, Washington, DC.

Wang, X. & Lu, J. 1999. The application of IC (integrated circuit) cards in efficient irrigation management system. *China Water Resources,* 436: 40 (in Chinese).

Ward, D.R. 2002. *Water Wars, Drought, Flood, Folly and the Politics of Thirst.* Riverhead Books – Penguin Putnam, Inc., New York.

Williamson, O.E. 2000. The new institutional economics: taking stock, looking ahead. *Journal of Economic Literature*, 38(September): 595–613.

Zekri, S. & Easter, K.W. 2007. Water reforms in developing countries: management transfer, private operators and water markets. *Water Policy,* 96: 573–589.

Chapter 4

Governance and the challenges of water and food security

Rob de Loë[1] & Henning Bjornlund[2]
[1]*University of Waterloo, Ontario, Canada*
[2]*University of South Australia, Adelaide, Australia; University of Lethbridge, Alberta, Canada*

Abstract

Water security exists when adequate water of appropriate quality is available for both human and environmental needs. Food security exists when all people have access to sufficient safe, nutritious food to meet their needs. Water security and food security are intimately interrelated because attempts to increase food security can compromise water security, for instance, when irrigation leads to degraded aquatic ecosystems. In an important shift in thinking, there now is widespread acceptance of the fact that technological solutions alone will not solve food and water security problems. Instead, it is necessary also to address the root causes of food and water insecurity, many of which relate to mismanagement, inappropriate policies, and weak *governance.* The concept of governance refers to the ways in which decisions that affect water are made, who is involved in making those decisions, and how power is distributed in society. This chapter explores links between water security, food security, and governance. A case study of the water governance in the Murray–Darling Basin is used to illustrate important considerations, including the changing roles of individuals, communities, and corporations in water governance in this region.

Keywords: Food security; Water governance; Water security

1 Introduction

Food security is a concern for individuals, communities, countries, and regions. According to the Food and Agriculture Organization (FAO) of the United Nations, food security exists "when all people, at all times, have physical, social and economic access to sufficient, safe and nutritious food to meet their dietary needs and food preferences for an active and healthy life" (FAO, n.d.). Given that an estimated 923 million people around the world suffered from chronic food deficits in 2007, a number that grew by 80 million since 1990–92 (FAO, 2008), food security clearly is a pressing humanitarian crisis.

Water quality and availability are important contributors to food security (United Nations World Water Assessment Programme (WWAP, 2006)). For instance, clean water and sanitation contribute directly to food security through their impacts on human health. Contaminated drinking water, inadequate sanitation, and poor hygiene result in a cumulative "disease burden" in the form of malnutrition, injuries, disabilities, and reduced capabilities (Prüss-Üstün *et al.*, 2008). An important outcome of this disease burden is a decreased ability of people to grow their own food and to care for themselves and their families (World Health Organization (WHO, 2009).

This book focuses on sustainable irrigation. Thus, the links between water availability and food security are a particular concern. Approximately 17% of the world's cropland is irrigated, and this land produces over one-third of the food and fibre grown globally (Oster & Wichelns, 2003). Reflecting the increasing importance of irrigation, there has been a fivefold increase in irrigated cropland relative to the beginning of the 20th century (Rosegrant *et al.*, 2002). Further expansion in the amount of land irrigated is likely in the future, with the bulk occurring in developing countries where population growth and demand for increased food production are the strongest (WWAP, 2003). Unfortunately, serious concerns exist regarding the ability of the world's water resources to meet new demands for irrigation. Globally, water withdrawals for irrigation are estimated at approximately 70% of total withdrawals, making agriculture by far the largest user of water on the planet (WWAP, 2003). Water withdrawals for domestic purposes, industry, and livestock are expected to increase by at least 50% by 2025, significantly reducing the amount of water available for expanded irrigation (Rosegrant *et al.*, 2002). Much of the expected future demand for water in support of agriculture will come from transboundary basins (Appelgren & Klohn, 1997), meaning that countries depending on these water resources must find ways to share them equitably. At the same time, concern for the impacts of water withdrawal on ecosystems is increasing (Dyson *et al.*, 2003). Thus, water resources that in previous decades might have been considered available for human use are now being seen as necessary for the maintenance of aquatic ecosystems.

The issues discussed in the previous paragraphs are thoroughly documented in numerous studies and reports produced by a host of organizations, including the FAO, the WWAP, and the WHO. In other words, there is nothing novel in the assertion that food security is determined in part by water quality and availability. However, there are two important themes in recent literature that do represent a shift in thinking, and which are pertinent to this book. First, there is growing concern for the extent to which attempts to increase food security can compromise *water security*. The impacts of agricultural production on water quality and quantity have long been recognized in the water field. To illustrate, it is well understood that in many parts of the world irrigation practices have resulted in lowered groundwater levels, salinization of soils, reduced water quality, alteration and regulation of natural watercourses, and reduced surface water availability (Dougherty & Mody,

1995; Rosegrant *et al.*, 2002). Nonetheless, clearer recognition of this concern in the food security literature is an important development. Second, and more importantly, now there is widespread acceptance of the fact that technological solutions alone will not solve food and water security problems. Instead, it is necessary also to address the root causes of food and water insecurity, many of which relate to mismanagement, inappropriate policies, and weak *governance*.

Governance is recognized as a critical concern in the contemporary water literature, but the term itself has many meanings. Easter and McCann (this volume) adopt the perspective of Williamson (2000) and view governance as the realm in which decisions are made in the context of higher level institutions that define the rules of the game. In this chapter, we adopt a perspective on governance that reflects broader considerations relating to the ways in which decisions are made, who is involved in making those decisions and how power is distributed in society. Governance, we argue, is a key determinant of the extent to which water security exists, and water security shapes, and is influenced by, food security. These relationships are explored in this chapter, with particular reference to irrigation. In the next section, we explore the nature of water security and link this concept to irrigation. This is followed by an overview of the importance of governance for water security. A case study—irrigation and water management in Australia—is used to illustrate the importance of these relationships.

2 Water security and irrigation

The concept of "water security" entered the language of the water field in the early 1990s, in the context of water scarcity and the perceived need to increase water supplies (Witter & Whiteford, 1999). Water security is now a common concern in the international water literature, in large part because of the Ministerial Declaration of The Hague on Water Security in the 21st century in 2000, and the Global Water Partnership's (GWP, 2000) promotion of the term. In its numerous variants, the concept of water security is relatively straightforward. When countries, regions, communities, industries, or individuals lack access to adequate supplies of water of appropriate quality, they lack water security. Thus, poverty and water security are inextricably linked in many places (WWAP, 2003). At the same time, though, water security is also linked to environmental quality. The GWP's (2000) definition of water security, for example, suggests that human needs must be balanced against the needs of aquatic ecosystems. For purposes of this chapter, we suggest that water security exists when sufficient water of good quality is available for social, economic, and cultural uses while, at the same time, adequate water is available to sustain and enhance important ecosystem functions (de Loë *et al.*, 2007).

As suggested above, water security directly shapes food security. This reflects the fact that in most ways in which water is used in agriculture, it is not substitutable. For instance, animals require water of acceptable quality in order to live, and plants require sufficient water of appropriate quality to grow and produce a harvest. When water is not available for livestock in natural watercourses, it must be provided by the producer. During times of water scarcity, livestock producers may be able to truck water to their animals or move their animals to water sources. However, their ability to do so depends, of course, on the availability of access rights to water supplies, or land with water supplies, and the availability of equipment and resources needed to move water to animals, or animals to water.

In plant agriculture, options are even more constrained. Where precipitation is not sufficient to meet the moisture needs of crops, irrigation can be a way to increase food security (WWAP, 2006). However, whether irrigation is feasible in a particular place depends on a host of considerations discussed in this book, including land and soil that are irrigable; water supplies in surface water bodies or aquifers; physical and legal access to those supplies; adequate infrastructure, technologies and skills; and appropriate institutions.

As these examples demonstrate, the link between food security and water security is quite complex. Irrigation can increase food security, but it can also increase food *insecurity* if the water resources themselves are not secure, for example, where groundwater resources are over-allocated, or where the source of water is a transboundary watershed in which competition occurs between sovereign jurisdictions sharing the resource (WWAP, 2003). Simultaneously, as a sector that withdraws and consumes vast amounts of water from surface water and groundwater sources, and also contributes to degraded land and water quality, irrigation itself is an activity that can compromise both water and food security. Therefore, it is essential that irrigation be viewed in a larger context, and that these kinds of links and relationships be addressed through effective governance.

3 Governance and water security

Modernization of irrigation systems is occurring throughout the world through upgrading of infrastructure and through improvements in water use efficiency and productivity in agriculture. This is being accomplished with technologies such as drip irrigation and laser-levelling, and through procedures such as improved irrigation scheduling—measures that permit application of water in the optimum quantity and timing for crop development (Oster & Wichelns, 2003). Measures such as these are an important part of any strategy designed to use water more efficiently and to adapt to scarcity. However, as pointed out by both Young and McColl and Nicol *et al.* (Chapters 5 and 13 in this book), increased water use efficiency can also threaten water security as it influences return flow and thereby supply for downstream users and in-stream needs. Hence, technology alone will not be enough to address the kinds of water security challenges that are being faced today. Therefore, considerable attention is also being paid to adapting institutional arrangements and irrigation policies in order to change the focus from developing new water supplies to a more holistic approach that considers agricultural productivity and market conditions. Reform of water allocation systems and increased use of economic instruments is seen as a priority in these efforts (Bjornlund, 2003; Dinar & Mody, 2004; Young & McColl, 2008).

Technological and institutional improvements within the irrigation sector are important contributors to its long-term viability. To the extent that they lead to more efficient water use and reduced impacts on land and water resources, they may also contribute to water security. However, improved practices and institutions within the sector will not be sufficient on their own to ensure water security, and thus the continued viability of the irrigation sector. As Bazzani *et al.* (2005) have noted, "Irrigated agriculture is currently confronted with many issues relating to its role within the overall management of water at [the] basin level". This suggests that what is also required is attention to the broader factors that shape water security in any particular place. The kinds of factors that are important can be revealed by considering the specific problem of water allocation.

Water allocation systems are the rules and procedures through which access to water for both consumptive and non-consumptive uses is determined (Tarlock, 1988). By establishing the availability and priority of access to water resources for consumptive uses such as irrigation, cities and manufacturing, and for non-consumptive uses such as hydropower, recreation and environmental protection, water allocation systems influence economic productivity, social and cultural well-being, and ecosystem quality (Gleick, 1998; Warner et al., 2006). Effective, efficient, and equitable water allocation systems are critical to maintaining and enhancing environmental quality, economic productivity, and social well-being. This is especially true in regions where local economies are strongly dependent on irrigated agriculture (Bjornlund, 2004). The socioeconomic, cultural and ecological implications of water allocation are amplified when water resources become scarce due to population growth, climate change, and changes in societal preferences. In the context of scarcity—whether created by societal or natural processes—water allocation systems can increase or decrease water security.

Table 1 draws attention to the broader factors that shape water security in the specific context of water allocation. Considerations such as these determine the extent to which sufficient water of good quality is available for social, economic and cultural uses while, at the same time, adequate water is available to sustain and enhance important ecosystem functions. For example, the need to find a balance between environmental and socio-economic concerns is strongly evident in concern for *ecosystem protection*, on the one hand, and *economic production*, on the other (Table 1). Water security is enhanced when water allocation systems recognize the need to provide water in appropriate amounts and at appropriate times, to support environmental needs, and when they incorporate mechanisms for monitoring and enforcing environmental water allocations, and are sufficiently flexible to permit the use of new ecological knowledge. Equally, water security is promoted by allocation rules that are clear and stable, by systems that provide information needed to make economically sound decisions, and by mechanisms that promote allocative efficiency.

The extent to which many of the concerns identified in Table 1 can be addressed successfully depends strongly on the effectiveness of governance. As noted

Table 1: Linking water allocation and water security.

Broad water security consideration	Specific concerns pertinent to water allocation
Ecosystem protection	• Environmental water allocation • Monitoring and enforcement for ecosystem protection • Creation and incorporation of ecological knowledge
Economic production	• Clear and stable allocation rules • Water allocation and related information to make economically sound decisions • Ability to reallocate water between users, sectors, and/or regions
Equity and participation	• Equity • Sustained and meaningful stakeholder and public participation • Mechanisms to address potential conflicts at different scales
Integration	• Integration between groundwater and surface water resources • Integration between water quality and quantity • Integration between land use planning and water allocation
Water conservation	• Conservation-related charges • Reallocation of water to more efficient and less consumptive uses • Incorporation of water conservation practices
Climate variability and change	• Investments to understand impacts of climate variability and change • Development and application of adaptation strategies
Transboundary sensitivity	• Coordination of water allocation systems across political boundaries • Respect for state sovereignty

Source: de Loë, R., Varghese, J., Ferreyra, C. & Kreutzwiser, R. 2007. *Water Allocation and Water Security in Canada: Initiating a Policy Dialogue for the 21st Century,* Guelph Water Management Group, University of Guelph, Guelph, ON, Canada.

in the introduction, governance is concerned with the ways in which decisions are made, who is involved in making those decisions, and how power is distributed in society. There are countless ways in which governance can occur in relation to water resources. One useful strategy for distinguishing among them is to consider the role of state actors versus non-state actors. Historically, the state has been the dominant, if not the only, actor in many countries. Non-state actors such as individual water users, firms, non-government organizations, and multilateral organizations are more recent additions to the cast. This phenomenon is often referred to as a shift from *governments* to *governance* (Pahl-Wostl *et al.*, 2008; Sonnenfeld & Mol, 2002; Kaika, 2003). It reflects the fact that governments no longer are, and in many cases cannot be, the sole source of environmental decision-making authority (Bryant & Wilson, 1998; Plummer & Armitage, 2007). The growing importance of non-state actors in environmental governance reflects a host of considerations including limitations on the capacity of government agencies; pressure from citizens for a greater role in decision–making; acknowledgement of the increasing complexity of environmental management, and thus the need for more minds and different kinds of knowledge; and pressure from international funding agencies (Pahl-Wostl, 2004; de Loë & Kreutzwiser, 2006; Armitage, 2008; Bruch, 2003).

Water allocation traditionally has been the domain of the state. Governments commonly have constitutional responsibility for water, and thus hold the authority to define the rules under which water is allocated. Even in countries where the state is no longer the dominant actor in water allocation, it most likely was and actively divested itself of that responsibility. Chile is a well-known example of this phenomenon. Through the 1981 National Water Code and the 1988 Constitution, Chile's government transformed its role in water governance (Bjornlund & McKay, 2002; Saleth & Dinar, 2000). Legally and practically, water is now private property in Chile, independent of land. Under the Water Code and the Constitution, rights to use water are determined through the market, and allocated within and across sectors. Different rights exist for consumptive and non-consumptive users (such as irrigation and power generation, respectively). Conflicts are resolved by water users associations and the courts. State intervention is minimal. Governance for water allocation in Chile is thus widely distributed among a range of state and non-state actors.

The broadening of water governance beyond governments is a global trend. In countries such as Chile and Australia (discussed below), market-based approaches have been adopted wholeheartedly albeit from very different perspectives. However, in other places governance involves a wide range of non-state actors, but the market does not play a significant role. For instance, in Canada, provincial governments are primarily responsible for water allocation, and each has devised its own system. In Western Canada, modified prior appropriation systems are used, while in Eastern Canada, systems rooted in the doctrine of riparian rights dominate (except in Quebec, which has a separate legal tradition of civil law) (de Loë *et al.*, 2007). Only in the western province of Alberta does a water market exist. Provisions exist under the province's Water Act that permits temporary and permanent trades, but the market is relatively underdeveloped, and remains closely supervised by the government. In general, the economic instruments that are available are not well supported by the irrigation industry (Bjornlund *et al.*, 2007).

These concerns can all be drawn together by considering again the case of irrigation. In most places where it is practised, irrigation is a key water user. As noted earlier, it is estimated that irrigated agriculture accounts for approximately 70% of global water withdrawals (WWAP, 2003). In regions where irrigation relies on surface water bodies, it commonly is in conflict with new demands for water that have emerged, notably protection of aquatic ecosystems. At the same time, as noted earlier, in many parts of the world where irrigation is important, water has been over-allocated, soils have been salinized, water quality has been compromised, and natural watercourses and aquatic ecosystems have been disrupted. The broadening of water governance beyond governments in many countries has meant that concerns such as these are taken seriously—if for no other reason than the fact that citizens and environmental groups are no longer prepared to leave decision-making about something as fundamentally important as water to governments. As a result, in many jurisdictions the courts are being asked to resolve disputes, or to force governments to take specific actions regarding pollution, protection of ecosystems, and other concerns.

These trends have significant implications for the irrigation sector, specifically, and for water management, generally. Where the needs of irrigation may have been the primary (if not only) concern of water managers, they now must be balanced against broader concerns that exist at the basin level, including water security. In seeking this balance in regions where irrigation is a significant water user, the following kinds of issues must be addressed:

- How can demands for water from other sources, such as growing cities and demands for protection and enhancement of aquatic ecosystems, be met in catchments where extensive irrigation development exists?
- What is the appropriate role of the state relative to non-state actors (individual water users, corporations, and non-government organizations) in water allocation?
- Can effective water governance promote water security?

4 Case study: Irrigation, water security, and governance in the Murray–Darling Basin

Australia's Murray–Darling Basin covers approximately 1 million km^2, or just over 14% of the country's total land area (Australian Bureau of Statistics, 2008). Five jurisdictions share the basin: four states (New South Wales, Victoria, Queensland and South Australia) and the Australian Capital Territory (ACT). As a case study for this chapter, the Murray–Darling Basin is ideal. The basin is home to approximately 2 million people, some 40% living in urban areas with 25,000 or more people, including the largest city Canberra (population 350,000) (Australian Bureau of Statistics, 2008). At the same time, the Murray–Darling Basin is a critical part of Australia's agricultural economy. In 2005–06, crop and livestock production in the basin accounted for 39% of the total value of agricultural production in Australia (Australian Bureau of Statistics, 2008). Importantly, 65% of Australia's irrigated land is located in the basin. Irrigated agriculture is reliant on water from the basin's rivers, and drought and over-allocation of water have placed enormous pressure on this resource. Reflecting these challenges, a series of important water reforms have taken place since the mid-1990s, with the specific goal of enhancing water security in the basin. These reforms have transformed water governance.

Historically, the various states sharing the Murray–Darling Basin have had very different interests and concerns. In South Australia, the main concern was originally for navigation, while for the upstream states of New South Wales and Victoria, irrigation, settlement and food production were the chief concerns (Clark & Renard, 1974). These divergent priorities caused conflicts in early transboundary water management agreements. A convergence of interests around managing and exploiting water for irrigation and food production occurred during the last century in all states. Environmental and water quality issues did not become apparent until the 1960s. Not until 1973 were the first steps taken to include these issues in the terms of reference of the then River Murray Commission with interim powers given in 1976. Not until the amendment of the River Murray Water Agreement in 1982 were they formally included in the commission's mandate (Pigram, 1986; Doyle & Kellow, 1995).

Under Australia's constitution, states have primary responsibility for water management. All the basin states created water allocation systems based on licences or entitlements for a defined period of time; in reality, these were treated as being in perpetuity. Entitlements were eventually defined in volumetric terms. However, access to water was defined each season as a percentage of the entitlement depending on availability. Agricultural production varied between states, and thus allocation policies developed in each state reflecting the needs of the major water user groups. Consequently, very different water allocation systems emerged. Up to the late 1990s, the situation among the three key basin states of New South Wales, Victoria, and South Australia was as follows:

- New South Wales was dominated by the rice industry and other broad acre crops, with land being relatively plentiful in supply. These industries have the ability to expand or contract their irrigated area on a seasonal basis depending on water availability. Thus, large volumes of unused or partly used entitlements were created and seasonal allocations were set very high, leaving very little in the storages at the end of each season as security for next year's allocations; as a result, supply security was low. Prior to the current drought, long-term seasonal allocations have been at 70% of entitlements.
- In Victoria, the major water user was the dairy industry. This industry has a higher level of need for supply security to support significant investments in permanent pastures, dairy herds, and equipment. Very low seasonal allocations could have long-term financial effect on their production. Hence, Victoria's allocation policy has been more conservative than that of New South Wales. Historically and prior to the recent drought, the average allocation level has been 160% of entitlements.

- South Australia is dominated by horticulture and viticulture with a very high dependence on a secure supply of water, and the potential for significant long-term losses if water is unavailable. As a result, the South Australian government has been very conservative in issuing new licences. South Australia was the first state in the basin to recognize over-allocation issues, by placing a moratorium on the issuing of new entitlements in 1969, following the 1967/68 drought, and by reducing existing entitlements according to actual use in 1969, 1976, and 1979 (Curd & Schonfeldt, 1990).

Special provisions for joint management of the Murray–Darling Basin have existed since 1915 when the *River Murray Waters* Act was created and the River Murray Commission was formed to oversee construction of works on the river. New institutions were created in the 1980s as concern shifted from irrigation to broader concerns associated with water sharing. Thus, in 1983, the Murray–Darling Basin Act was passed, and in 1985, the Murray–Darling Basin Commission (MDBC) was formed. The *Water Act 2007* put in place a framework for environmental protection, monitoring, enforcement, water sharing, and trading, and created the independent Murray–Darling Basin Authority (MDBA) to replace the MDBC.

Up until the 1990s, water governance in Australia was very state centered, and embodied a top–down approach to planning and decision-making. All water allocation decisions were made by the state governments, and irrigation districts were government owned and operated. This started to change when the Council of Australian Governments (COAG) introduced a new water policy framework in 1994 (COAG, 1994). The new framework included six important elements: (1) water prices should reflect volumetric consumption at a rate based on full cost recovery, including environmental cost; (2) water entitlements should be separated from the property rights in land and associated with clear specifications of ownership, transferability, reliability, and, where appropriate, quality; (3) trading in water entitlements should be encouraged to ensure that water is used to maximize its contribution to national income and welfare within social, physical, and ecological constraints of catchments; (4) irrigators should be given greater influence over the management of irrigation areas by transferring operational responsibilities relating to districts to local bodies; (5) communities should be involved in natural resource management issues and education programs should be implemented to improve the ability of the community to participate in the decision-making processes; and (6) specific entitlements should be given to the environment, acknowledging it as a legitimate user of water.

The COAG policy put in place the foundations for a new approach to water governance in Australia. However, pressures on water resources continued to increase throughout the 1990s and into the early part of the 21st century. For example, between 1894 and 1993, annual discharge at the mouth of the Murray–Darling system has ranged from 1.626 million dam^3 to 54.168 million dam^3 with a median of 8.49 million dam^3 (Murray–Darling Basin Commission, 2005). However, due to the drought inflows into the basin in the years since 1993 have been considerably reduced, and showed no signs of improving. In the 2 years ending in August 2008, inflows into the Murray system were 3.54 million dam^3, just over 50% of the previous 2-year minimum of 6.8 million dam^3 in 1943–45 (Murray–Darling Basin Commission, 2008). Recognizing the magnitude of the crisis, the COAG increased the pace of reform through the 2004 National Water Initiative (NWI), which expanded water trading and focused attention on the environmental and economic needs of the Murray–Darling Basin (McNamara, 2007). The NWI was designed to address three main concerns: (1) uncertainty over the long-term access to water continued to hamper investments in higher valued and more efficient production systems; (2) current water market arrangements were preventing markets from reaching their full potential; and (3) serious concerns existed regarding the ability of the reforms flowing from the 1994 framework to secure adequate environmental flows, and to permit adaptive management to ensure the health of riverine ecosystems.

The NWI was agreed to by all states and the Commonwealth. Thus, it provides Australia's overarching water policy framework (Hussey & Dovers, 2006). Importantly, the NWI must be seen as part of a movement away from entirely top–down water planning that began with the 1994 COAG reforms, and culminated most recently with the *Water Act 2007.* The result has been a dramatically different approach to water governance than the one that existed in the years prior to 1994. The state and Commonwealth governments remain critical actors in water governance. However, governance has become much more widely distributed beyond the state through the water reform process. Three examples illustrate how individuals, communities, and corporations now play important roles in water governance in Australia.

- First, under the NWI, states are obliged to return all catchments to sustainable levels of extractions (Council of Australian Governments, 2004). This will be accomplished through the creation of water sharing plans, which are regulatory documents. Catchment management organizations in each catchment are required to define environmental needs. They must define the amount of water that is available for consumption (referred to as the consumptive pool), and then determine how the consumption pool will be shared among existing entitlement holders (Smiley *et al.*).
- Second, one of the MDBA's responsibilities under the *Water Act 2007* is creation of a single, consistent, and integrated basin plan. Public involvement and consultation in this process are required. The Act establishes a basin community committee that advises the MDBA on plan development and implementation (Commonwealth of Australia).
- Third, a key part of the water reform process has been the creation of water markets and the transfer of irrigation districts to private corporations. As a result, critical decisions about how water will be used are being made by companies and individuals engaging in the water market.

Australia's approach to water governance is widely acknowledged to be proactive and forward looking. However, implementation has not been problem-free.

Water planning processes reliant on community involvement have been met with considerable scepticism within many communities, particularly in NSW (Kuehne). The lack of clarity regarding the nature of community involvement in water planning led to frustration, confusion, and uncertainty among many irrigators, and has caused "consultation fatigue". Another critical challenge to the legitimacy of distributed governance is the fact that implementation of water sharing plans developed at the community level has been postponed. Indeed, under the current drought, New South Wales has simply ceased the implementation of the water sharing plans. Citizens involved in the development of some plans also have been frustrated by the fact that their work and recommendations can be overruled by the minister in charge of water, as has occurred in New South Wales (Smiley *et al.*).

Another important problem is the tension that exists regarding the balance between statutory and market-based approaches. Under the NWI, the need to provide water for environmental and other public benefit outcomes is a statutory requirement, and this requirement is the main instrument designed to achieve sustainable levels of extraction. Unfortunately, the drought has intensified since 2004, and statutory processes have largely failed to achieve sustainable levels of extraction and secure water for key environmental sites (Young & McColl, 2008; Thoyer, 2006; Connell, 2007). Therefore, markets have increasingly been promoted as the means for government to purchase water from willing sellers to ensure water for the environment and reduce over-allocation (Loch *et al.*, 2009). This represents a very different approach to "public" involvement than was envisioned under the NWI.

5 Discussion and conclusions

The central argument in this chapter is that governance is a key determinant of water security. In other words, we argue that the extent to which demands for water for human uses such as cities and irrigation can be met while ensuring water for the protection and enhancement of aquatic ecosystems depends strongly on the processes and institutions through which decisions are made. Clearly technology plays a role. For instance, dams and reservoirs are a key part of water management systems in most regions where large volumes of water are needed reliably to meet demands for irrigation. Similarly, new technologies, such as more efficient sprinkler systems, are an important part of any strategy to increase the efficiency of water use. Nonetheless, in countries around the world, it has become clear that technology alone cannot create water security. This relates in large part to the extent to which poor management practices, weak institutions, and wasteful practices contribute to the water crisis faced in many jurisdictions (WWAP, 2003).

Water management in Australia's Murray–Darling Basin, the case study examined in this chapter, amply supports this argument. For example, water storage facilities are an important part of the water management system in the basin, and billions of dollars currently are being invested in new facilities and improvements to existing infrastructure. However, inflows into the basin have reached such extremely low levels during the last few years that there simply is not enough water to fill reservoirs. To illustrate, in August 2008, active storage in the Murray system was only 20% of capacity (Murray–Darling Basin Commission, 2008). Reforms to systems for water governance in Australia, driven in large measure by circumstances in the Murray–Darling Basin, have therefore been the focus of efforts to improve water security in Australia since the mid-1990s. As discussed in this chapter, these reforms have included a near-total transformation of water allocation systems and a fundamental realignment of the roles of governments (Commonwealth and state) and non-state actors (including corporations, water users, and citizens). Major changes nationwide include a consistent national policy, the establishment of water markets, and a greatly expanded role for citizens. In the Murray–Darling Basin, additional reforms include a basin management agency with considerable authority for transboundary water management, and specific legal rules and procedures for water sharing among the basin states.

Relative to the links between water allocation and water security identified in Table 1, the importance of governance in water reforms in the basin is clear:

- *Ecosystem protection* has become a central concern at both the state and the basin level. Within states, provisions are established for environmental water allocation through the guidance of the NWI. In the Murray–Darling Basin, additional provisions exist, including an Environmental Water Holder, basin-scale planning and funds to support the rehabilitation of priority sites. Unfortunately, the severity of the drought has led to the suspension of key planning processes designed to ensure water for the environment in New South Wales. Instead, attention is shifting to using markets to recover water for the environment, which represents new terrain from the viewpoint of water governance.
- From the viewpoint of *economic production*, rules for water allocation are now much clearer than previously. Individual water users now have full information on the value of water through the markets, and reallocation of water among users, sectors, and regions is possible. Whether the new rules are *stable* is debatable. For instance, the pace of reform since 1994 has been extreme, and, as demonstrated by the example of environmental water allocation in New South Wales, the drought emergency puts into question whether even the newest reforms have produced a robust system capable of dealing with current let alone future uncertainty (Young & McColl).
- In terms of *equity and participation*, there is no question that new opportunities for sustained and meaningful citizen involvement have been created. However, government continues to be the dominant actor in the basin, as demonstrated by the sweeping powers that have been assigned to the Commonwealth government and the MDBA through the *Water Act 2007*. As noted in the context of environmental water allocation, whether public involvement can be considered *meaningful* in circumstances where decisions made by catchment management organizations can be set aside by ministerial fiat is questionable (Smiley *et al.*). This concern

intensifies if, as is starting to seem likely, the primary vehicle for "public" involvement in water governance becomes participation in water markets.

- *Integration* is a priority in water reforms that have occurred in the basin, as witnessed by efforts to bring groundwater and surface water management under the purview of the MDBA, and the nature and extent of attention that is given to water quality as well as water quantity concerns in the Murray–Darling Basin Agreement. Similarly, through institutions such as the revised Murray–Darling Basin Agreement, Australia's governments have demonstrated strong recognition of the extent to which water allocation, land use planning, and economic development are integrated.
- *Water conservation* practices are an important way of linking water allocation systems and water security. In addition to using funds to construct new infrastructure, the Commonwealth has provided financial resources to upgrade irrigation systems and to promote more efficient water use in irrigation. At the same time, irrigation efficiency has increased dramatically as low-value producers have left the sector and been replaced by farmers growing high-value crops and using water more efficiently. However, there is increasing concern over the net effect of increased water use efficiency on water security as long as water entitlements are defined as extractive gross rights rather than net use rights. Under such systems, most water saved is not really conserved as it is used to expand irrigation or sold to other consumptive users. Furthermore, many such savings actually result in losses to the system as the saved water would otherwise have returned to the system (see Chapters 5 and 13 of this book).
- *Climate variability and change* is taken extremely seriously in Australia. The Commonwealth government sees the current drought not only as an indicator of climate change but also as a likely future (Garnaut, 2008). Hence, climate change has been a major driver of the water reforms. To illustrate the seriousness of the concern, the Commonwealth's recent investigation into the impacts of climate change suggests that in the Murray–Darling Basin, a 1°C increase in temperature is expected to produce a 15% decrease in streamflow in the basin. The report suggests that irrigated agriculture in the basin could effectively cease before the end of the century due to reductions in river flows, increases in drought frequency, and increased evaporation from water storage reservoirs (Garnaut, 2008).
- Finally, in the context of *transboundary sensitivity*, the case examined in this chapter demonstrates an extremely distinctive approach, relative to practically any transboundary situation, to dealing with jurisdictional boundaries. While states in basin have not lost any of their constitutional authority over water, they most certainly have delegated considerable responsibility to an organization created under a Commonwealth statute (the MDBA). Comparable examples of jurisdictions with such clear authority over water handing over so much authority to a transboundary agency are rare.

This analysis demonstrates unequivocally that water security is directly shaped by the quality and effectiveness of governance. Who is involved in decision-making, how they are involved, and how it is decided who is involved have been key concerns in the water reforms that have occurred in Australia, generally, and specifically in the Murray–Darling Basin. Paradoxically, however, this case also demonstrates that it is possible for neither technology nor governance to be sufficient to the challenge. The water crisis in Australia's Murray–Darling Basin is so severe that no amount of water reform may be adequate to the task. Further, reforms have been proposed by some analysts with the goal of making water allocation systems in the basin more robust and "future proof" (Young & McColl, 2008). Whether these will be adequate will be the subject of much debate in the years to come. However, one lesson that is absolutely clear from the case of water allocation in the Murray–Darling Basin is that jurisdictions facing water scarcity, and even those that are relatively well-off at this time, should avoid at all costs encountering similar circumstances. Effective governance can promote water security – but not under any and all circumstances.

References

Appelgren, B. & Klohn, W. 1997. Management of transboundary water resources for water security: principles, approaches and state practice. *Natural Resources Forum,* 21(2): 91–100.

Armitage, D. 2008. Governance and the commons in a multi-level world. *International Journal of the Commons,* 2(1): 7–32.

Australian Bureau of Statistics. 2008. *Water and the Murray-Darling Basin. A Statistical Profile 2000-01 to 2005-06.* ABS Catalogue No. 4610.0.55.007, Australian Bureau of Statistics, Canberra, Australia.

Bazzani, G.M., Di Pasquale, S., Gallerani, V., Morganti, S., Raggi, M. & Viaggi, D. 2005. The sustainability of irrigated agricultural systems under the Water Framework Directive: first results. *Environmental Modelling & Software,* 20(2): 165–175.

Bjornlund, H. & McKay, J. 2002. Aspects of water markets for developing countries: experiences from Australia, Chile, and the US. *Environment and Development Economics,* 7: 769–795.

Bjornlund, H. 2003. Efficient water market mechanisms to cope with water scarcity. *Water Resources Development,* 19(4): 553–567.

Bjornlund, H. 2004. Formal and informal water markets: drivers of sustainable rural communities? *Water Resources Research,* 40: 1–12.

Bjornlund, H., Nicol, L. & Klein, K.K. 2007. Challenges in implementing economic instruments to manage irrigation water on farms in southern Alberta. *Agricultural Water Management,* 92(3):131–141.

Bruch, C. 2003. Role of public participation and access to information in the management of transboundary watercourses (Chapter 2). In: Nakayama, M. (Ed.), *International Waters in Southern Africa.* United Nations University Press, Tokyo and New York, pp. 38–70.

Bryant, R. & Wilson, G. 1998. Rethinking environmental management. *Progress in Human Geography,* 22(3): 321–343.

Clark, S.D. & Renard, I.A. 1974. Constitutional, legal and administrative problems. In: Frith, H.J. & Sawer, G. (Eds.), *The Murray Waters. Man, Nature and the River System.* Angus and Robertson, Sydney, Australia, pp. 263–281.

Commonwealth of Australia, State of New South Wales, State of Victoria, State of Queensland, State of South Australia, and Australian Capital Territory, Agreement on Murray-Darling Basin Reform. Made in Sydney on 3 July 2008.

Connell, D. 2007. *Water Politics in the Murray-Darling Basin.* The Federation Press, Annandale, NSW, Australia.

Council of Australian Governments. 1994. *Council of Australian Governments Water Reform Framework.* Australian Government Publishing Service, Canberra, Australia.

Council of Australian Governments. 2004. *Intergovernmental Agreement on a National Water Initiative.*

Curd, V. & Schonfeldt, C. 1990. Water transfers: the South Australian experience. In: Pigram, J.J. & Hooper, B. (Eds.) *Transferability of Water Entitlements.* Centre for Water Policy Research, Armidale, NSW, Australia, pp. 201–211.

de Loë, R.C. & Kreutzwiser, R. 2006. Challenging the status quo: the evolution of water governance in Canada. In: Bakker, K. (Ed.), *Eau Canada: The Future of Canadian Water Governance.* University of British Columbia Press, Vancouver, Canada, pp. 85–103.

de Loë, R., Varghese, J., Ferreyra, C. & Kreutzwiser, R. 2007. *Water Allocation and Water Security in Canada: Initiating a Policy Dialogue for the 21st Century,* Guelph Water Management Group, University of Guelph, Guelph, ON, Canada.

Dinar, A. & Mody, J. 2004. Irrigation water management policies: allocation and pricing principles and implementation experience. *Natural Resources Forum,* 28: 112–122.

Dougherty, T.C. & Mody, J. 1995. *Environmental Impact Assessment of Irrigation and Drainage Projects.* Irrigation and Drainage Paper No. 53, Food and Agriculture Organization of the United Nations, Rome.

Doyle, T.J. & Kellow, A.J. 1995. *The Murray Waters. Man, Nature and the River System.* McMillan Education Australia, Melbourne, Australia.

Dyson, M., Bergkamp, G. & Scanlon, J. 2003. *Flow: The Essentials of Environmental Flows.* IUCN—The World Conservation Union, Gland, Switzerland.

Easter, K.W. & McCann, L.M.J. 2009. Sustainable water projects: the task of economic instruments and supporting institutions. In: Bjornlund, H. (Ed.), *Incentive and Instruments for Sustainable Irrigation.* WIT Press, Southampton, UK.

Food and Agriculture Organization of the United Nations, 2009. Glossary. http://km.fao.org/fsn/resources/glossary/en/.

Food and Agriculture Organization of the United Nations. 2008. *The State of Food Insecurity in the World 2008.* Food and Agriculture Organization of the United Nations, Rome.

Garnaut, R. 2008. *The Garnaut Climate Change Review: Final Report.* Cambridge University Press, Port Melbourne, Cambridge, New York and Melbourne.

Gleick, P.H. 1998. Water in crisis: paths to sustainable water use. *Ecological Applications,* 8(3): 571–579.

Global Water Partnership. 2000. Towards water security: a framework for action. Global Water Partnership.

Hussey, K. & Dovers, S. 2006. Trajectories in Australian water policy. *Journal of Contemporary Water Research and Education*, 135: 36–50.

Kaika, M. 2003. The water framework directive: a new directive for a changing social, political and economic European framework. *European Planning Studies,* 11(3): 299–316.

Kuehne, G. and Bjornlund, H. (2006): “Custodians” or “Investors” – classifying irrigators in Australia’s Namoi Valley. In Lorenzini, G. and Brebbia, C.A. eds. *Sustainable Irrigation Management, Technologies and Policies.* Southampton: WIT Press, 225–236.

Loch, A., Bjornlund, H & Kuehne, G. 2009. *Purchasing Water to Create Sustainable Systems: Where Does This Leave the Regulatory Approach?* Environment Research Event.

McNamara, P. 2007. Murray Darling Basin: Drought, Over-Allocation, Conservation. *Trust News Australia*, 14–16.

Murray–Darling Basin Commission. 2005. Murray-Darling Basin e-Resources 2005. http://www.mdbc.gov.au/subs/eResource_book/index.htm. Accessed 25 August 2009.

Murray–Darling Basin Commission. 2008. *MDBC Murray System Drought Update.* Issue 15, September 2008, Murray-Darling Basin Commission, Canberra, Australia.

Oster, J.D. & Wichelns, D. 2003. Economic and agronomic strategies to achieve sustainable irrigation. *Irrigation Science*, 22(3–4): 107–120.

Pahl-Wostl, C. 2004. The implications of complexity for integrated resources management. In: Pahl-Wostl, C., Schmidt, S. & Jakeman, T. (Eds.), *Complexity and Integrated Resources Management.* International Environmental Modelling and Software Society, Osnabrück, Germany.

Pahl-Wostl, C., Gupta, J. & Petry, D. 2008. Governance and the global water system: a theoretical exploration. *Global Governance*, 14: 419–435.

Pigram, J.J. 1986. *Issues in the Management of Australia's Water Resources.* Longman Cheshire, Melbourne, Australia.

Plummer, R. & Armitage, D. 2007. Crossing boundaries, crossing scales: the evolution of environment and resource co-management. *Geography Compass,* 1(4): 834–849.

Prüss-Üstün, A., Bos, R., Gore, F. & Bartram, J. 2008. *Safer Water, Better Health: Costs, Benefits and Sustainability of Interventions to Protect and Promote Health.* World Health Organization, Geneva.

Rosegrant, M.W., Cai, X. & Cline, S.A. 2002. *Global Water Outlook to 2025: Averting an Impending Crisis.* International Food Policy Research Institute and International Water Management Institute, Washington, DC and Colombo, Sri Lanka.

Saleth, R.M. & Dinar, A. 2000. Institutional changes in global water sector: trends, patterns, and implications. *Water Policy,* 2: 175–199.

Smiley, S., de Loë, R.C. & Kreutzwiser, R.D. Appropriate public involvement in local environmental governance: a framework and case study. *Society & Natural Resources,* (in press).

Sonnenfeld, D.A. & Mol, A.P.J. 2002. Globalization and the transformation of environmental governance. *American Behavioral Scientist,* 45(9): 1318–1339.

Tarlock, A.D. 1988. *Law of Water Rights and Resources.* Clark Boardman, New York.

Thoyer, S. 2006. How to reallocate water rights when environmental goals conflict with existing entitlements. *International Journal of Sustainable Development,* 9(2): 122–137.

United Nations World Water Assessment Programme. 2003. *Water for People, Water for Life: The United Nations World Water Development Report.* UNESCO Publishing, New York.

United Nations World Water Assessment Programme. 2006. *Water: A Shared Responsibility.* UNESCO and Berghahn Books, Barcelona, Spain.

Warner, J.F., Bindraban, P.S. & Van Keulen, H. 2006. Introduction: water for food and ecosystems: how to cut which pie? *Water Resources Development,* 22(1): 3–13.

Williamson, O.E. 2000. The new institutional economics: taking stock, looking ahead. *Journal of Economic Literature,* 38(2): 595–613.

Witter, S.G. & Whiteford, S. 1999. Water security: the issues and policy challenges. *International Review of Comparative Public Policy,* 11: 1–25.

World Health Organization, 2009. Health impacts of the global food security crisis. http://www.who.int/food_crisis/en/. Accessed 26 August 2009.

Young, M. & McColl, J. 2008. *A Future-Proofed Basin: A New Water Management Regime for the Murray-Darling Basin.* University of Adelaide, Adelaide, South Australia.

Young, M. & McColl, J. 2009. A robust framework for the allocation of water in an ever-changing world (Chapter 4). In: Bjornlund, H. (Ed.) *Incentive and Instruments for Sustainable Irrigation.* WIT Press, Southampton, UK.

Chapter 5

A robust framework for the allocation of water in an ever-changing world

Michael D. Young[1] & James C. McColl[2]
[1]University of Adelaide, Australia
[2]CSIRO Sustainable Ecosystems, Adelaide, Australia

Abstract

In an ever-changing world with an increasing demand for water, and with the emerging possibility that climate regimes may change, there is a need to put in place arrangements to enable the water resource system to adapt and support itself no matter what the future may bring. Irrigation could be expected to come and go, rivers and the associated environment must be expected to change and the expectations and attitudes of communities and societies will also change.

This chapter focuses on the nature of water allocation regimes that are designed to cope with adverse shifts in water supply, with changing input and product prices, with changing demands for water, and to facilitate adjustment through the adoption of new technologies and management approaches. In essence, we seek guidelines for the development of an allocation and associated administrative regime that is robust—robust in the sense that the regime can be described confidently as likely to withstand the test of time, no matter what climatic and economic future arrives. A robust allocation regime for a regulated river system is defined that first allocates water to enable the maintenance of functioning river sufficient to allow water to be extracted for use, then allocates the available water as unit shares between consumption and environment entitlements. A key feature is the need to give careful attention to water accounting arrangements to ensure hydrological integrity of the system and to arrangements that encourage smart and innovative management of environmental water.

Keywords: Water allocation; Climate change; Robust systems; Water accounting; Environmental entitlements; Water markets

1 Introduction

Change is an enduring process without beginning and end (McColl & Young, 2005). There is nothing more constant than change. Prices change, technologies change, populations change and they all want access to water. Moreover, in recent times, the challenge of coping with such pressures for change and adjustment has been exacerbated by the emergence of the possibility that climatic regimes may change. The need to put in place arrangements to enable the water resource system to adapt and support itself no matter what the future may bring is becoming increasingly urgent. Irrigation could be expected to come and go, rivers and the associated environment must be expected to change, and the expectations and attitudes of communities and societies will also change.

Water users around the world and in Australia, in particular, are learning about the severity and speed with which adverse climate change can impact on water supplies. The top half of Figure 1 presents rainfall data from the weather station that best describe the sudden shift in rainfall that occurred in 1994 in the region that supplies Perth's dams. The regime appears to have become drier, much drier. The bottom half of Figure 1 illustrates the dramatic impact that this drier regime has on stream inflow into the dams.

As a crude general rule, for every 1% reduction in mean rainfall there is at least a 3% reduction in stream inflow. Careful inspection of Figure 1 reveals that since 1974 this region has never received that which was thought from historical records to be its mean stream inflow.

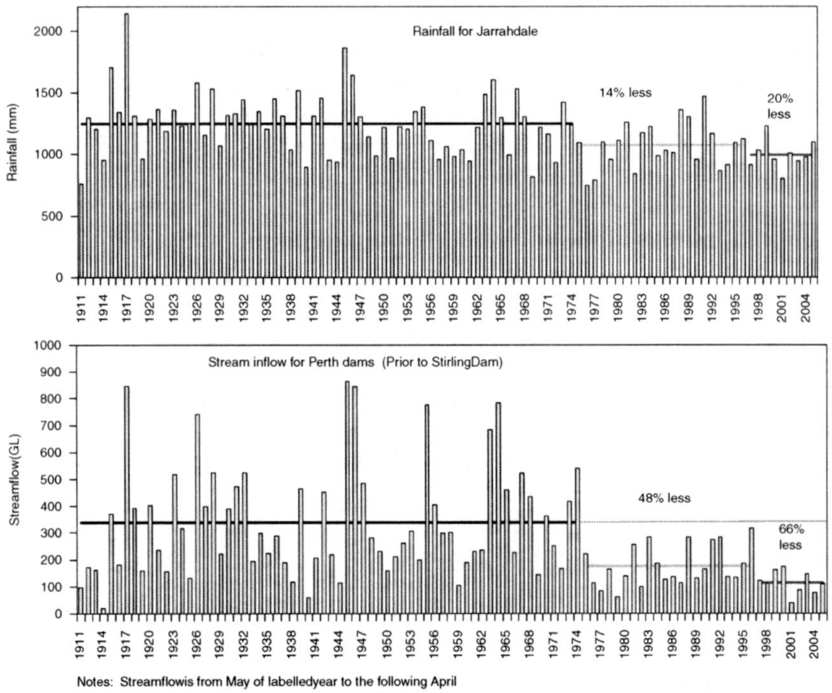

Figure 1: Change in rainfall and inflows into water storages near Perth, Western Australia.

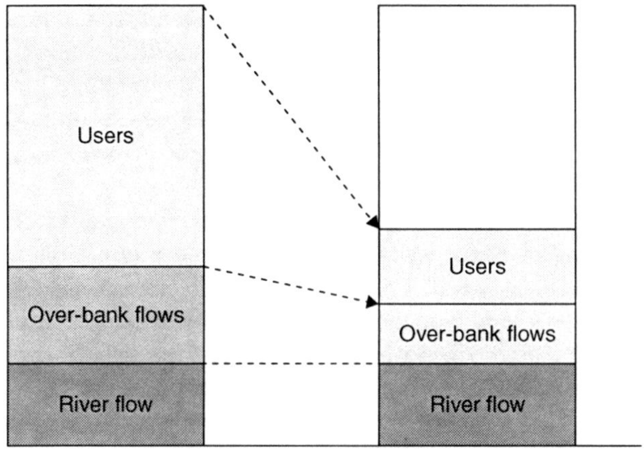

Figure 2: Illustration of the impact of a shift to a drier regime on the amount of water that can be made available for use in a regulated river system.

In fact, the situation can be worse than this, especially in regulated river systems that are used to convey water to users and facilitate navigation. In such systems, a relatively large proportion of water is needed simply to maintain the river in a regulated state. As a result, and as shown in Figure 2, when the mean amount of water flowing into a system is halved, the amount available for use can be reduced by two-thirds. Changes of this magnitude continue to surprise policy makers.

In this chapter, we focus on the nature of water allocation regimes that are designed to cope with adverse shifts in water supply, with changing input and product prices, with changing demands for water, and facilitate adjustment through the adoption of new technologies and management approaches. In essence, we seek guidelines for the development of an allocation and associated administrative regime that is robust—robust in the sense that the regime can be described confidently as likely to withstand the test of time, no matter what climatic and economic future arrives (Young & McColl, 2005).

2 Outcomes, principles and concepts

The outcomes sought are a suite of arrangements that will maintain the health of a river and the communities that depend on them. One of the main features of robust systems is that they facilitate autonomous adjustment as and when change occurs.

Robust systems have a demonstrated ability to *recover gracefully* from the whole range of exceptional inputs and situations in a given environment. They tend to be one step below *bulletproof*. They endure without the need to change their foundations. They inspire confidence and are capable of producing efficient, politically acceptable outcomes in an ever-changing world.

Robustness is a concept that has received little attention in the literature. When searching for guidelines to assist with the design of robust allocation systems, however, several Nobel Prize-winning principles are worth careful consideration (Young & McColl, 2002].

The first of these principles is the *Tinbergen principle* which observes that if we are interested in dynamically efficient resource use through time, at least as many instruments should be used as there are objectives. That is, we should never try to use the same instrument to achieve, for example, both equity and efficiency in resource use. In Australia, appreciation of the importance of this concept has resulted in the unbundling of most water entitlements. In such a system, the arrangements used to define water entitlements are kept separate from those used to define allocations and both of these are kept separate from those used to manage the environmental impacts associated with the application of water on land. In the most sophisticated of these systems, separate instruments are also used to manage water delivery and salinity.

A related but equally important concept is *Mundell's assignment principle* which requires that each instrument be assigned to the objective where it generates the most leverage and that once assigned to that objective the instrument not be used to help achieve any other objective. In Australia's River Murray system, for example, allocations—the amount of water that an entitlement holder may use in any one period—are managed separately from the entitlement—the instrument used to define each entitlement holder's share of the any water likely to be made available for use. Allocation policy is used to enable efficient resource use and entitlements used to ensure that the allocation processes used are equitable.

A third concept of equal importance is the *Coase theorem* which stresses the importance of keeping market transaction costs to a minimum so that adjustment is possible in a timely manner and all participants in a water allocation system are able to negotiate with one another. If the cost of trading seasonal allocations is low and trades can be completed quickly, then water use will be much more efficient than it otherwise would be.

3 Defining a robust allocation regime for a regulated river system

For any river system, it is common to begin by stating that the first allocation priority should be to allocate water to the environment. In practice, we think this approach is inadequate in that it fails to differentiate between the water needed to maintain a functioning river and that needed to enable water to flow periodically over a bank and into the surrounding landscape. The first priority should be the allocation of sufficient water to support the maintenance of a regulated river at

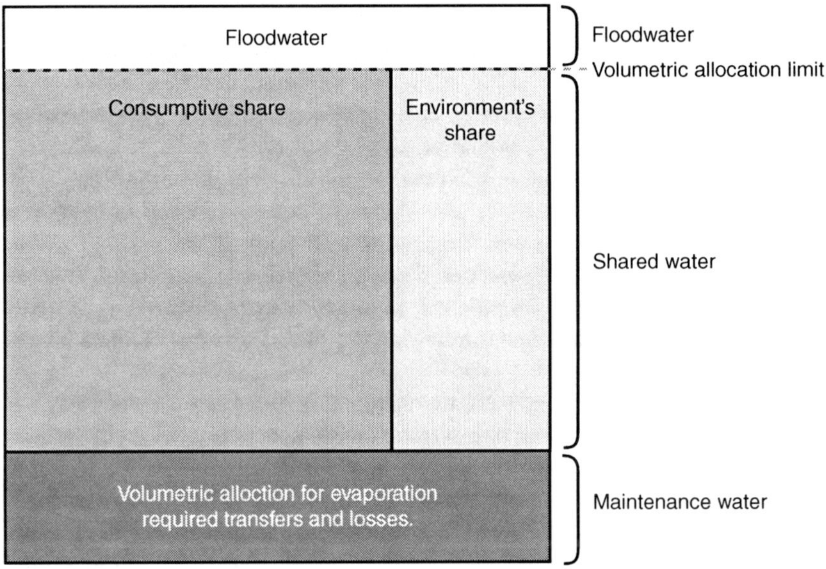

Figure 3: A robust water allocation regime. The first priority is to allocate the minimum amount of water necessary to maintain the system. The second priority is to allocate water to the environment for over-bank use and to consumptive users.

the minimum level required to enable water to be extracted for any use, including the over-bank environment.

As summarised in Figure 3, we propose that the next tranche of water should be shared between consumptive users and the environment. Differing from convention, we recommend this approach because it ensures that the environment always gets some water. With formal shares issued to this environment water and defined in the same manner as for consumption, it is administratively impossible to allocate water to consumptive water users without allocating some water to the environment for use beyond the river.

The third and last tranche of water is floodwater. As a general rule, formal entitlements to floodwater should not be issued. The main reason for this is to ensure that no one is liable for the damage caused by floodwater.

When assigning entitlements to water in the shared tranche, we like to describe this tranche as a variable pool of water. Drawing on the well-established and well-tested mechanisms used to define ownership in companies, entitlements should be defined as unit shares rather than proportional shares. The main reason is that this makes it possible to surrender shares and to move shares from one system to another without having to change every share in the share register. In essence, the entitlement regime should be fully specified in the sense that the sum of all entitlements to water that is available for sharing totals 100%.

4 System bulk allocations

When designing a water allocation regime, it is important to differentiate between the bulk allocation rules designed to manage allocations at the system level and those designed to manage use by individuals.

The simplest bulk allocation system that one can imagine establishes a single entitlement pool and allocates unit shares to those interested in holding an entitlement to access any water allocated to that pool. If the costs of trading entitlement shares and/or allocations made to shareholders are zero, this will produce efficient water use. In practice, however, adjustment costs exist and in most systems it is more efficient to partition the shared water pool into a number of priority classes.

Usually, the highest priority entitlement pool is given to providing water for critical human needs, and then one or more entitlement pools of varying reliability are established.

In the River Murray system, for example, special rules have been put in place to ensure that once conveyance or minimum maintenance water requirements have been met, the next priority is to supply the minimum amount of water needed to meet critical human needs. Once critical human needs have been met, two bulk entitlement pools are normally established—a high security pool and a general and lower security pool. The general rule is that the maximum volume that may be assigned to the high security pool is defined and once this volume has been assigned to that pool and allocations made to high security entitlement holders, allocations are then made to those who hold shares in general and low security pools.

In the more sophisticated systems, arrangements are now being put in place to ensure that the high security pool is climate change adjusted so that all water users receive an early warning about shifts in water availability. Advocating this approach, Young and McColl propose that the volume of the high security pool should be adjusted using a long-run moving average of inflows into the system (Young & McColl, 2008a).

As mentioned earlier, it is necessary to differentiate between water needed to cover evaporative and in-stream losses and that needed for over-bank purposes. In many management systems, this water is simply described as environmental water. In practice, however, the two types of water are quite different and it is misleading to describe them using the same term. In regulated river systems, in-stream water is needed by all consumptive users so that water can be conveyed to them. We call this water "maintenance water". In Australia's Murray–Darling System, this is now being described as conveyance water.

An interesting new policy development has been a recent change in the way it is planned to define the environment's right to access shared water. Australian water legislation typically has stated that the environment should have first claim to any water allocations and once this need has been met, the remaining water may be allocated. In practice, however, this approach has not worked. As a result, governments are now moving to a regime where entitlements to water for over-bank and other non-conveyance purposes is being defined by issuing the environment with a formal entitlement to a share of the water allocated to each security pool. In fully or over-allocated systems, the approach taken is purchase entitlements from existing users and transfer ownership to an environmental water holder. In the Murray–Darling Basin, for example, A$3.1 billion has been allocated for the purchase of water entitlements and A$5.8 billion for investment in water savings projects that will enable the transfer of existing entitlements to a Commonwealth Environmental Water Holder (Water Act, Parliament of Australia). It is proposed that this program will be implemented over the 10 years to 2016/17.

5 Individual interests

Having established a framework for the system bulk allocation regime, we can now turn to the nature of arrangements necessary to ensure that individual water use is efficient, that allocations are equitable and that the arrangements can cope with sudden as well as gradual changes. Young and McColl have argued the case for the definition of all entitlements using unit shares recorded on a central register (Young & McColl, 2005; 2002).

The case for issuing shares rather than volumetric entitlements is predicated primarily on the need to make it clear that investment in water use is risky and that there can be no guarantee that a specific volume of water will always be available. Sharing systems also put in place a suite of arrangements that make it clear how the available water is to be shared. In essence, the requirement is that allocations be made to a defined pool and every shareholder is then entitled on a pro-rata basis to a proportion of those allocations to the pool. No guarantee is made about reliability.

As stated earlier, it is important that shares are defined as unit rather than proportional shares so that additional shares may be issued and shares transferred to another part of the system without the need to change every share in the system whenever, for example, administrative boundaries are changed or investments that increase water availability are made.

Similarly, administrative efficiency can be increased by establishing and using a share register to define ownership. In these systems, the register defines who owns what. The transfer of ownership from one person to another is then implemented by contracting to change the register. This Torrens title system or property registration, as it is called, is much more reliable than any system that defines ownership by issuing a certificate. When a share register is used to define entitlement, arrangements that allow the registration of mortgages, etc. are easily added to the system. The entire system can be managed electronically and efficient settlement procedures established.

6 Water accounting

One of the key features of any robust water entitlement and allocation system is the need to give careful attention to water accounting arrangements that ensure that all transactions and allocation arrangements have hydrological integrity. Allocation systems that have hydrological integrity are characterised by the fact that allocation policies take full account of system interconnectivity—especially ground–surface water interconnectivity, processes that intercept flows which otherwise would have reached the river, and water use practice changes that reduce the return of "inefficiently" used water to the river system. The essential rule is that whenever one person or process is allowed to take more water, someone else in the system understands that as a result of this action they will have to take less water (Young & McColl, 2003).

The most common error in Australian water accounting has been the failure to properly account for return flows and to define entitlements in gross rather than net terms. Under a gross entitlement allocation system, increases in water use efficiency reduce return flows to the system (Young & McColl, 2003; Crase, 2008). Under a gross allocation system, an entitlement holder can improve the efficiency of water use and, by reducing the volume of water that is returned to the system, increase the total volume of water they use, and decrease the volume available for others to use. In contrast, net entitlement systems only allow water users to keep real savings that result from a reduction in the volume of water that has been returning to the system either via a surface drain or through accession to groundwater.

Robust water entitlement systems also have to account for connectivity between surface and groundwater systems. The first step in such a process is to define all water within a short distance of a river as river water rather than groundwater. Beyond that area, the entitlement regime needs to specify the relationship between the river and the groundwater system in a way that changes in one resource influence the management of the other. Taking this concept to its logical conclusion, Young and McColl recommend that groundwater entitlement systems that contribute to river flow be defined so that the river holds shares in the groundwater allocation system (Young & McColl, 2008b). The river should also be required to contribute a minimum volume of maintenance water to the groundwater system before any allocations are made to those who hold shares in the groundwater system. Where this cannot be done, an alternative equally robust approach is to require that groundwater depth be kept within a narrow range.

The reverse applies in the case of groundwater systems that gain water or are recharged by a river. In this case, robustness requires that the nature of this relationship be fully specified by allocating shares in the river system to the groundwater system.

7 Interception

A related accounting challenge is the question of how best to account for the interception of water before it reaches a river or storage. Examples of processes that intercept water include the expansion of plantation forestry, the construction of farm dams and the construction of levy banks designed to prevent overland flows from reaching a river.

Conceptually, the impact of interception is similar to that of a shift to a drier climatic regime. This form of water use, however, can be managed, and if it is, investment will be more efficient than it otherwise would be (Young, 2009). To illustrate this point, consider the likely impact of the introduction of a carbon emissions trading scheme on water users. Stimulated by the opportunity to earn emissions trading permits, landholders have an incentive to plant trees and as a result decrease the amount of water available for use by irrigators. The plantation forestry sector gets access to water that irrigators were previously entitled to use. To manage this process, plantation forestry and all other forms of land-use change that intercept water need to be bought into the allocation system.

As the amount of water taken by most interception processes is not and cannot be metered, an alternative approach is needed to account for the impact of these processes. In recognition of the proposition that "it is better to be approximately right than comprehensively wrong", Young and McColl (2009) recommend that an assessment be made of the expected long-term impact of an interception process, and interceptors be required to surrender entitlements whose expected water use would be the amount that is expected to be intercepted. Recognising that under such an arrangement, the interception process is protected from the adverse effects of climate change, they go on to recommend that the amount surrendered could include an adverse climate change premium.

8 Allocation management

In many systems, there is pressure on system managers to make allocations in a manner that gives confidence to water users. One of the more serious mistakes made in Australia in recent times was to assume that the supposed "drought" would end and it could not get worse. In the Murray–Darling Basin, for example, until very recently it was assumed that the minimum volume of water available for use would always be greater than the sum of the minimum inflow received in each month. As summarised in Table 1, in 2006/07 inflows into the River Murray system were less than the minimum monthly inflow for 11 months in succession with the result that allocations had to be reduced after they had been announced. As a result, it has now been decided to shift to a more robust arrangement that only allows allocations to be made when delivery can be guaranteed. The result is much more robust and makes it clear to users that they are responsible for managing supply risks and uncertainty.

9 Facilitating adjustment in an ever-changing world

A related issue is the question of how to proceed down the pathway of water reform in a way that facilitates change and adjustment in an ever-changing world. In a dramatic change to policy and as part of the development of a National

Table 1: Recorded minimum inflows into the River Murray system before 2006/07 and in 2006/07.

Months	Previous lowest monthly inflow before 2006		2006 inflow	Previous minimum inflow (%)
June	110	GL in 1967	110	100
July	150	GL in 1967	130	87
August	130	GL in 1902	100	77
Sept	180	GL in 1902	120	67
October	140	GL in 1914	80	57
November	60	GL in 1914	70	117
December	60	GL in 1982	60	100
January	50	GL in 1983	50	100
February	60	GL in 2003	50	83
March	50	GL in 1915	50	100
April	70	GL in 1923	40	57
May	80	GL in 1902	110	138
Total	1140		970	

Source: Murray–Darling Basin Commission, personal communication, 2008.

Competition Policy in 1993/94, Australian governments agreed to separate water licences from land and make these licences—which became known as entitlements—tradable.

The first step in the process was to introduce arrangements to enable water licences or water rights to be held separately for land titles, thus enabling water to be traded separately from land. Having achieved that goal, as part of a National Water Initiative agreed in 2004, it was decided that to improve the efficiency of water trading arrangements, water licences, or water rights as irrigators prefer to call them, should be unbundled. Under such an arrangement, entitlements, allocations and permissions to apply water to land could be managed separately. Importantly, as part of these reforms and in the search for a more robust way to define entitlements, it was decided that entitlements should be defined as shares and that no implicit reliability guarantees should be given. It was also made clear that the responsibility for managing climatic risks and uncertainty should be assigned to entitlement holders and not to government.

In a similar manner, allocations are to be managed separately, and in the search for a robust management system, they are being registered on accounts which, like bank accounts, are accessible over the Internet. Under such an arrangement, whenever a trade occurs, one account is debited and simultaneously

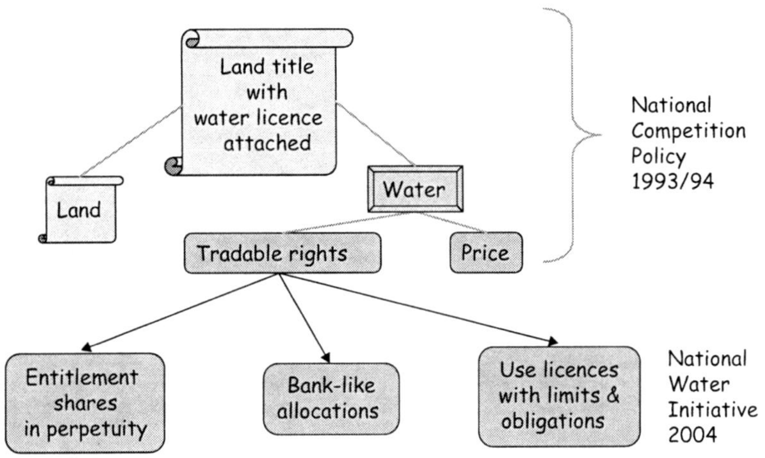

Figure 4: Evolution of water reform in Australia and the progressive separation of water licences from, and subsequent unbundling into, separate components.

another account is credited. As illustrated in Figure 4, the third element of this unbundled arrangement is an approval to apply water to land which is usually subject to an array of local conditions.

Under such an arrangement, entitlement and allocation markets have rapidly emerged and grown in sophistication, efficiency and robustness.

In the early stages in the development of these markets, water allocations were only tradable within a season. In essence, the rule was to use it, or sell it, as you cannot save it (Young & McColl, 2007a). It soon became apparent, however, that if an efficient outcome is to be achieved then markets have to allow the management of water supply both within and between seasons (Brennan, 2007a; 2007b).

Having established these foundations for the development of entitlement markets and allocation markets, the last step in the development of efficient markets is to introduce a suite of rules that prevent markets from being used to protect water users from competition. Whilst progress has been slower than one would expect, it has now become clear that this requires a suite of arrangements that

- Assign ownership of entitlements to individual water users and not to irrigation districts so that any entitlement holder is free to choose whether to sell their entitlement either within or outside an irrigation district;
- Regulate the setting of fees and charges associated with supply and delivery contracts so that any person who decides to sell their water outside an irrigation district is not charged an inequitable fee for deciding to do this (ACCC, 2006; 2008);
- Require trades to be completed within a reasonable amount of time;
- Prevent districts from placing limits on the proportion of entitlements and/or the quantity of water that can be traded out of a district.

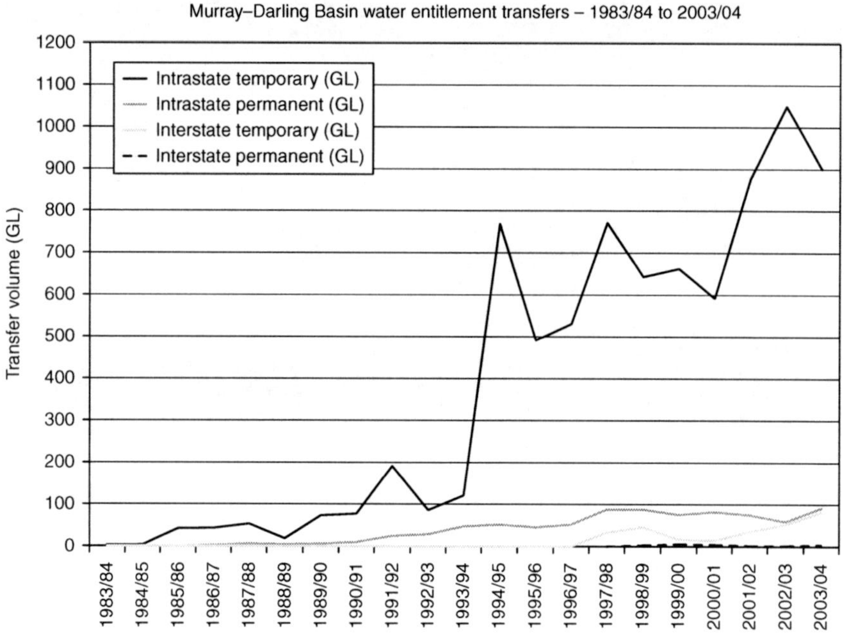

Figure 5: Growth in water trading in the southern connected portion of the Murray–Darling Basin, 1983–2004.

In summary, under a robust arrangement it should be possible for the application of trading rules and processes associated with trades to be completed instantaneously for allocation trades and within two days for unencumbered entitlement trades (Young & McColl, 2008a).

10 Managing environmental water

In the allocation system set out, we make a clear distinction between conveyance or maintenance water and water needed for the maintenance of the environment on either side of a river. Differing from conventional practice, we have recommended that this latter form of water be defined in exactly the same way as all other shared water entitlements are defined. Under such an allocation regime, whenever an allocation is made to the shared water pool, some water must be allocated to the environment. Environmental managers must then decide whether to use this water, save it for future use or sell it in the market for allocation water.

In passing, we note that under such an arrangement the optimal environmental water management strategy may sometimes be to engage in what Australians describe as counter-cyclic trading. Counter-cyclic trading involves the sale of water to consumptive users in the peak of a dry period or a drought (long dry period)—when prices are high. This money can then be used subsequently to buy back a lot more water when water is cheaper. According to Kirby *et al.* (2006), in a regime where supplies are variable, this can be an extremely profitable strategy for an environmental manager to pursue.

In large river systems, like Australia's River Murray, one of the more interesting challenges to resolve is the question of whether to assign environmental entitlements to regional environmental management trusts or to hold these titles centrally. Having considered the question, we conclude that the most robust approach is a mixed one. A considerable proportion, say 70%, can be assigned to regional trusts and a much smaller amount needs to be held centrally to enable the pursuit of systemwide opportunities at the margin. This approach to the assignment of environmental water entitlements allows local catchment boards and communities to plan and manage with confidence. It can also be expected to encourage the development of smart and innovative environmental water managers.

11 Concluding comment

At the start of this chapter, we set out to specify the main characteristics of a water entitlement and allocation regime that could be confidently described as robust—robust in the sense that the regime could be expected to withstand the test of time and not require modification as conditions change. We emphasise that the need is to put in place arrangements to enable the water resource system to adapt and support itself no matter what the future may bring is becoming increasingly urgent. Irrigation could be expected to come and go, rivers and the associated environment must be expected to change and the expectations and attitudes of communities and societies will also change.

One of the main features of the model proposed is that regime gives first allocation to maintenance of the river, but not the over-bank watering priorities. In recognition that governments, while willing to commit to giving first priority to the environment, are not willing to do so when the system is under pressure, a much more secure sharing approach is recommended. Under such a regime, the property right is fully specified and commitments to make a proportion of water available for the environment must be honoured. As would happen in a natural system, if the system gets drier, the environment gets less water.

References

Crase, L. 2008. *Water Policy in Australia: The Impact of Change and Uncertainty.* Resources for the Future, Washington DC.

Brennan. D. 2007a. Managing water resource reliability through water storage markets. Paper presented to the Australian Agricultural and Resource Economics Society Conference, Queenstown, New Zealand, available at: www.myoung.net.au/water/policies/Brennan_Storage_Markets.pdf.

Brennan. D. 2007b. Missing markets for storage and their implications for spatial water markets. Paper presented to the Australian Agricultural and Resource Economics

Conference, Queenstown, New Zealand, available at: www.myoung.net.au/water/policies/Brennan_AARES07_storage_&_trade.pdf.

Australian Competition and Consumer Commission (ACCC). 2006. A regime for the calculation and implementation of exit, access and termination fees charged by irrigation water delivery businesses in the southern Murray–Darling Basin. Commonwealth of Australia.

Australian Competition and Consumer Commission (ACCC). 2008. Water market rules issues paper. Commonwealth of Australia.

Kirby, M., Qureshi, M.E., Mainuddin, M. & Dyack, B. 2006. Catchment behaviour and counter-cyclical water trade: an integrated model. *Natural Resource Modelling*, 19(4): 483–510.

McColl, J.C. & Young, M.D. 2005. Managing Change: Australian structural adjustment lessons for water. CSIRO Land and Water. Report no. 16/05.

Young, M.D. & McColl, J.C. 2002. Robust Separation: A search for a generic framework to simplify registration and trading of interests in natural resources. CSIRO Land and Water. Report Folio no. S/02/1578.

Young, M.D. & McColl, J.C. 2003. Robust reform: the case for a new water entitlement system for Australia. *Australian Economic Review*, 36(2): 225–234.

Young, M.D. & McColl, J.C. 2005. Defining tradable water entitlements and allocations: a robust system. *Canadian Water Resources Journal*, 30(1): 65–72.

Young, M.D. & McColl, J.C. 2007. Irrigation water: Use it or lose it because you can't save it. Droplet 6, 2007a, available at: www.myoung.net.au.

Young, M.D. & McColl, J.C. 2008a. *A Future-Proofed Basin: A New Water Management Regime for the Murray–Darling Basin*. The University of Adelaide, Adelaide.

Young, M.D. & McColl, J.C. 2008b. Grounding connectivity: Do rivers have aquifer rights? Droplet 13, 2008b, available at: www.myoung.net.au.

Young, M.D. 2009. The effects of water markets, water institutions and prices on the adoption of irrigation technology. In: Albiac, J. & Dinar, A. (Eds.), *The Management of Water Quality and Irrigation Technologies*. Earthscan, London.

Young, M.D. & McColl, J.C. 2009. Double trouble: the importance of accounting for and defining water entitlements consistent with hydrological realities. *Australian Journal of Agricultural and Resource Economics*, Volume 53, Number 1, January 2009.

Water Act 2007 (as amended), Parliament of Australia.

Part II: Case Studies

Chapter 6

Non-profit-maximising values and attitudes influencing irrigators' management response to new policy instruments

Geoff Kuehne[1] & Henning Bjornlund[1,2]
[1]*Centre for Regulation and Market Analysis, University of South Australia, Adelaide, Australia*
[2]*Department of Economics, University of Lethbridge, Lethbridge, Alberta, Canada*

Abstract

Undertaking effective and timely water reform in Australia is becoming more and more urgent to reverse increasingly apparent environmental damage and to achieve a sustainable irrigation industry. With farmers being Australia's largest users of water, the success of any reform depends on their involvement. It has therefore become much more important to understand all of the influences affecting the behaviour of farmers. With an ever more urgent need for action, it is now no longer acceptable to rely solely on an assumption of profit-maximising behaviour when devising programs and policies requiring their participation. The research reported in this chapter identifies and examines the non-profit-maximising influences on farmers' behaviour and shows how this insight can be used to more effectively tailor policies and programs so that they are more likely to achieve the desired result – which is, in this case, a more sustainable irrigation industry.

Keywords: Australia; Farmer decision-making; Farmers' values; Sustainable irrigation

1 Introduction

When governments and others wish to encourage irrigators to adopt activities and practices required for sustainable irrigation, they are essentially trying to change irrigators' behaviour and change the way in which they use and interact with water resources. Since farmers gain many benefits from irrigation other than just financial gain, they will take these benefits into account when making decisions in response to government policies designed to convince farmers to adopt practices and technologies that are likely to result in a more sustainable irrigation industry. It is important to understand that irrigators' management responses to these policies will be driven by their desire to maximise their total utility from irrigation, that is maximising all the benefits that they derive from irrigation, and not simply to maximise their financial profit. It is therefore essential for policy-makers to have a thorough understanding of the factors influencing irrigators' behaviour, other than just profit maximisation. This chapter contributes to this end by examining how the non-profit-maximising values that irrigators hold such as their concerns for family, profit, land, water and community influence their behaviours and decision-making.

The recognition among Australian policy-makers that non-profit-maximising values influence irrigators' decision-making has been slow to occur and has not previously had a significant influence on how policies and programs for water reform are designed or implemented. However, it does now seem to be slowly and tentatively incorporated into some new policies and instruments.

The context for the study reported in this chapter is the anticipated responses to new water sharing plans (WSPs) for groundwater licence holders within the Namoi Valley of New South Wales, Australia. The implications from the research findings are then discussed in relation to the recent offers by the Australian government to buy water from irrigators within the Murray–Darling Basin – the majority of Australia's irrigation area – aimed at achieving improved environmental sustainability (Wong, 2008).

2 Policy problem

The Australian water reform process has continued to intensify with the most recent development being the Rudd Government's $12.9 billion "Water for the Future" plan (Wong, 2008). This plan, in common with all water reform policies since the National Water Initiative of 2004 (COAG, 2004), relies on water markets as a mechanism to reallocate water to the environment, and also to more efficient and productive users of the water. From the time water markets (and their requirement for the separation of land and water titles) were first suggested as a method for water entitlement reallocation, irrigators were initially reluctant to embrace them (Bjornlund, 2006). This will need to change, however, as the Australian government intends to purchase $3.1 billion of water entitlements (approximately 15% of the average annual diversion by irrigators) over a 10-year period (Young & McColl, 2007).

Because the buy-back of entitlements has been implemented in a cautious and provisional fashion, with few purchases at the start of the period, increased pressure will be placed on purchases in subsequent years to achieve the overall required volumes. However, research by ACIL Tasman (2008) and Waterfind (2008) continues to suggest that irrigators will not be participating in water markets in the numbers required for the government to achieve its intended policy goals. At the current rate of market activity, the government will need to purchase all the water offered for sale over the next 14 years to reach its objective Waterfind, 2008). It is feared that this level of market intervention by the government could disrupt the market's ability to facilitate structural change for the benefit of current water users, as it has done previously (Bjornlund, 2008).

For the government to return sufficient water to the environment to achieve its sustainability targets, irrigators will therefore be expected to participate in the water markets to a greater extent than they have ever done previously. The implications of this study are that irrigators might be encouraged to sell water if they are approached in a way that takes account of their non-profit-maximising values, attitudes and goals. This understanding can be used to inform water reform, not only in the Namoi Valley or in other parts of Australia but also in other parts of the world. This study shows how the dimensions of family, land and profit, and to a lesser extent, water, community and lifestyle interact to influence the behaviour of farmers. Using the recommendations from this study, policy-makers could design policies and programs that better take account of the full range of influences on farmers' behaviour. In the context of the water entitlement buy-back scheme of the Australian government's "Water for the Future" plan (Wong, 2008), this might result in the increased uptake of water purchase offers among reluctant sellers and facilitate the move towards a more sustainable irrigation industry.

3 The study region——The Namoi Valley of New South Wales

The development of the Namoi Valley as an irrigation region has much to do with the uncertainty and unpredictability of rainfall that is the backdrop to much agricultural activity in Australia. Areas that are desperately short of rainfall, 1 year can be severely flooded the next. Irrigation is a response to this uncertainty of rainfall, and, depending on where it occurs and what type of irrigation is practised, it can reduce the risk involved with drought years.

Irrigation in the Namoi Valley was made possible through the building of Keepit Dam in 1961 and the Split Rock Dam in 1987 (Pigram, 2006). These dams were built to "conserve" the unruly flows of the Namoi River and its tributaries so that it could be used by agriculture, rather than be "lost" to the environment. Initially, at least, the government had to encourage potential irrigation farmers to recognise the promise presented by the water stored in these dams. Because this stored water was plentiful and obtained by irrigators at little cost, the use of groundwater was usually considered a less economically viable and consequently less preferred source of water.

However, groundwater became much more sought after during the drought of 1994–1996 when there was no allocation of water to irrigators from Keepit Dam. The government's response was to increase the groundwater allocations of the conjunctive licence holders (those with licences to extract both surface and groundwater) during the drought years. This resulted in irrigators, who had previously mostly relied on the more economical surface water, withdrawing large amounts of water from the aquifer (Kuehne & Bjornlund, 2006). It was this event that caused the first signs of severe depletion such as falling water tables and land subsidence (Namoi Groundwater Taskforce).

The unsustainable demands on the aquifer also occurred concurrently with the first steps taken towards water reform in the form of a cap to further diversions from the Murray–Darling system (NSW Government, 2000). Even though this was not to apply for some years in the Namoi Valley, it was a clear signal to irrigators that water would increase in value as its use was restricted.

The WSPs were developed to ensure sustainable groundwater extraction for the groundwater irrigators of the Namoi Valley (Murray–Darling Basin Ministerial Council, 1995). Achieving sustainable use of the resource has required cuts to irrigators' entitlements in all but 3 of the 13 hydrogeologically distinct zones [Namoi Groundwater Management Committee, 2001). In the most severely affected zone, this required a reduction of 94%. Because of the severity of reduction in water access, the implementation of this policy, even allowing for the potential adjustments likely to be achieved by water trading, was expected to lead to at least some of the licence holders experiencing significant economic pain.

The development of the WSPs was one of the key components of the Water Management Act 2000 (Powell *et al.*, 2003), and was aimed at satisfying one of the NSW government's unmet obligations remaining from the COAG agreement of February 1994 (COAG, 1994). The process firstly required that the amount of water needed to sustain the environment be defined, and then the implementation of a system of water access shares for the amount of water that was determined to be available for extractive use; a theme that was to continue with the signing of the National Water Initiative in 2005 (COAG, 2004).

In the Namoi Valley, it is this fundamental requirement of sustainable irrigation – that the needs of the environment are acknowledged first, with the remaining water shared for extractive purposes – that has been responsible for a conflict-ridden WSP development period (Kuehne & Bjornlund, 2006).

The reductions in entitlements to sustainable levels of extraction, specified in the WSPs (Aqualina, 2003), have caused much conflict between licence holders and the government. Friction was also caused among licence holders who had been using most or all of their entitlement and those who had been using little or none of their entitlement because the severity of entitlement reduction varied according to the licence holders' previous water use (Kuehne & Bjornlund, 2006).

4 Methods

The research methods were a mail-out survey to uncover and explore issues, followed by personal interviews to further develop the ideas gathered from the analysis of the mail-out survey. Based on this process, a phone survey instrument was designed to gather information from respondents about their (1) business, property and personal characteristics; (2) ratings on 50 value statements regarding family, land, water, community, lifestyle and profit and (3) land and water management actions and intentions. After refining the value statements with a focus group of agricultural extension officers, the survey instrument was piloted with 70 South Australian irrigators (Kuehne *et al.*, 2007).

The analysis of the telephone survey was conducted using factor analysis of the value statements. This was done to elicit the importance that farmers placed on the dimensions of family, land, water, community and profit, by forming value constructs (Kuehne & Bjornlund, 2008). The value statements were also cluster analysed to combine irrigators (those who responded indicating similar levels of importance for similar values) into distinct groups (Kuehne *et al.*, 2008). Investigations were undertaken to examine whether these value constructs led to different behaviours among irrigators and also whether these cases belonging to the cluster groups exhibited different forms of behaviour to each other by using cross tabulations and significance testing with the demographic, personal and the business and management actions and intentions data.

5 Findings

The mail-out survey was sent to 659 irrigators and resulted in a 23% response rate. The survey revealed that irrigators place varying degrees of importance on the dimensions of family, land, water, profit and community when deciding how to respond to the implementation of the WSP. Apart from identifying these dimensions, the analysis did not show how important the dimensions were, either in their own right, or relative to each other, or which combinations of these dimensions might be important for understanding or predicting their behaviour.

Following the mail-out survey, personal interviews were conducted with 17 licence holders to develop a greater understanding of the history and the problems of the region. As the mail-out survey found that farmers' intended responses to the WSPs were not often solely influenced by profit maximisation, the personal interviews were used to uncover and examine alternatives to profit maximisation. The interviews also provided the background information that allowed the questions and statements of the telephone interviews to be perceived as credible, relevant, and therefore worthy of consideration by participants.

Twelve of the 50 value statements from the telephone survey were used for factor analysis to form three value constructs incorporating the value dimensions of family, land, water, community and profit. The value constructs were described as *Succession*, *Caretaking* and *Commerce* (Kuehne *et al.*, 2008).

The *Succession* value construct included individual value statements that were related to the value of family. The *Caretaking* value construct contained value statements related to good land husbandry, responsible water use and a preparedness to accept that family might choose careers other than farming. The *Commerce* value construct contained value statements that were more clearly related to the achievement of profit, without an emphasis on family succession. These value constructs were not expected to be mutually exclusive but were expected to bemore or less important to all participants. Because the differences between those considering these value constructs to be important were not always straightforward, they illustrate the complexity of the influences on farmers' behaviour (see Figure 1). By focusing on the factor scores above the 40th percentile, Figure 1 shows the commonality between many of the cases occurring in each of the value constructs.

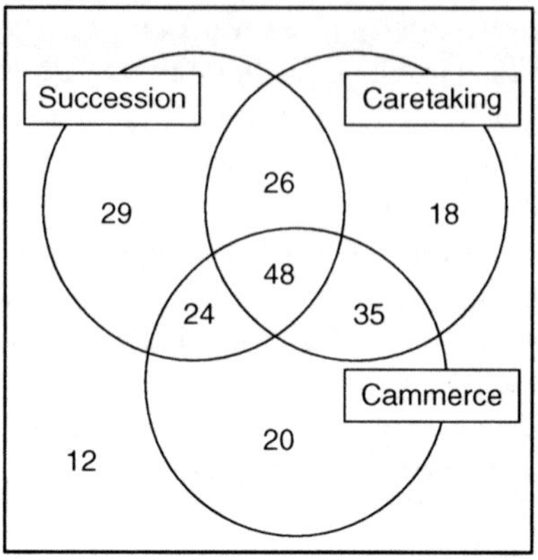

Figure 1: The value constructs are not mutually exclusive. Adapted from Kuehne, G. 2009. The influence of values and attitudes towards family, land, water, community, lifestyle, and profit on farmers' behaviour. Business, University of South Australia.

Factor analysis showed how the values that farmers considered to be important could be grouped into value constructs, how these value constructs overlapped and how they could be used to explain farmers' participation in water reform activities. However, to group these farmers according to the importance that they placed on these values, cluster analysis was used. Using 29 suitable value statements (where responses to the value statements were capable of differentiating individuals), a cluster analysis result was obtained with three plausible clusters: *Investors*, *Expanders* and *Traditionals*.

The results from the personal interviews were used to confirm that there was a credible match between the values that interviewees thought were important and their cluster group membership. Detailed examinations of the cluster groups using cross tabulations with the demographic, personal, and land and water management data were conducted. The dimensions of water, community and lifestyle were not found to be as important for defining the cluster groups, as were the dimensions of family, land and profit (see Figure 2). The *Investors* are defined most strongly by profit, the *Expanders* by a combination of profit and family, and the *Traditionals* by a combination of family and land (Kuehne *et al.*, 2008).

Participants varied in how much importance they placed on the value constructs; however, for ease of interpretation, only the strongest relationships with the value constructs are shown in Figure 2. The *Commerce* and the *Succession* value constructs have the most obvious interactions with the cluster groups, while the role of the *Caretaking* value construct is much less distinct.

Figure 2 is important for understanding how the value constructs interact with the cluster groups and for understanding the role of the most important value dimensions in each of the cluster groups. The principal characteristics of the

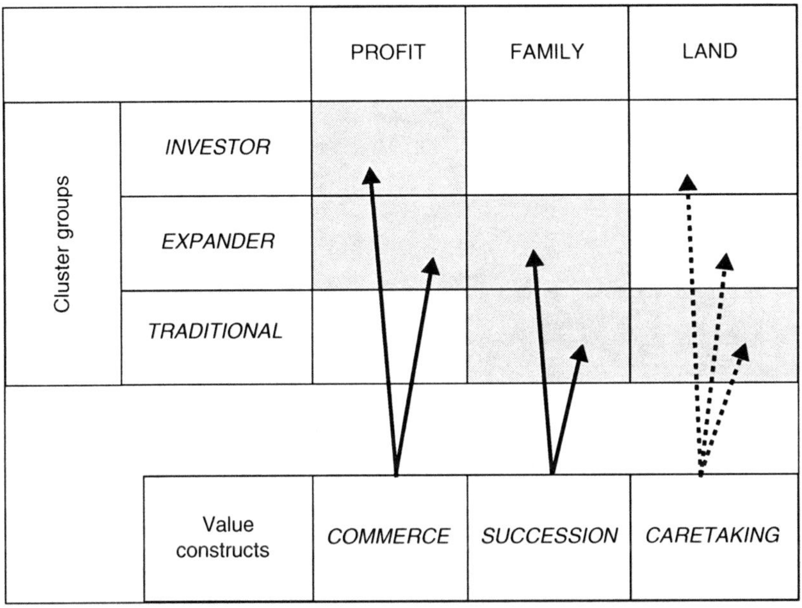

Figure 2: The value constructs are congruent with cluster group influences. Adapted from Kuehne, G. 2009. The influence of values and attitudes towards family, land, water, community, lifestyle, and profit on farmers' behaviour. Business, University of South Australia.

cluster groups (see Table 1) that have been developed from the value statements used for the cluster analysis provide a more detailed understanding of the cluster groups. The secondary characteristics have been derived from comparisons made between the demographic, personal, business, and land and water management data and the cluster groups.

Investors are most strongly influenced by the values that they hold regarding profit. They are not strongly influenced by the values that they hold towards family or land. They do not take a sentimental approach to farm ownership; rather, they are prepared to sell land when it represents a financially attractive opportunity. They are much younger and are more recent entrants to the farming industry. They grow more profitable crops and have more narrow profit-maximising goals, which are likely to include the realisation of a capital gain at the end of their farming career. They are therefore also far less likely to expect family succession. Reflecting these values, they have been more likely to sell or lease land and water in the past and are also more likely to anticipate doing so in the future (see Table 2).

Traditionals are strongly influenced by the values that they hold towards their family and land. *Traditionals* are likely to be established farmers with a traditional crop mix. They have a lesser drive towards generating profit and further business growth compared to the other groups. Nearly half of the *Traditionals*

Table 1: Irrigators with different values also have different demographic characteristics.

	Value characteristics	Demographic characteristics
Investor	Not family focused Profit focused Moderate attachment to land Favour water trading	Use more of their groundwater Have larger amounts of surface water More likely to grow cotton Younger, less experienced Lesser expectation of succession More educated
Traditional	Family focused Not focused on profit Strong attachment to land	Smaller amounts of groundwater More likely to grow pasture High expectation of succession
Expander	Family focused Profit focused Not attached to land. Community is important Favour water trading	Owners are older, more experienced Highest expectation of succession Less well educated

Source: Kuehne, G. 2009. The influence of values and attitudes towards family, land, water, community, lifestyle, and profit on farmers' behaviour. Business, University of South Australia.

Table 2: Values influence land and water management actions and intentions.

		Investor	*Traditional*	*Expander*
Water	Past	Have sold	Have sold	Have purchased
	Future	Will sell		Will purchase
Land	Past			Have purchased
	Future	Will sell	Will purchase	
Irrigated area	Past	Have reduced	Have reduced	Have maintained
	Future	Will reduce		Will increase

Source: Kuehne, G. 2009. The influence of values and attitudes towards family, land, water, community, lifestyle, and profit on farmers' behaviour. Business, University of South Australia.

expect to have family succession; however, nearly one quarter of that same group are also certain of not having succession. They do not see land as being a tradable commodity, instead it appears that it might satisfy other needs for them, which were not identified or explored by this study. They have a greater emphasis on dryland farming and tend to use their water for lower-value uses such as irrigating pastures for grazing. It could be that there are two types of farmers in this group: those with larger viable traditional family farms already with succession arranged and those who do not have any family succession in place but still reflect the underlying family values. The latter type would therefore have no ambition to develop their farm any further as they just want to remain on the farm as long as they can. At retirement, their farm would be sold to non-family members to support their retirement. Under current drought and policy conditions, the *Traditionals* seem to be in the process of shifting their enterprise balance so that they rely more heavily on dryland activities and less on irrigation.

Expanders consider profit to be important as well as family concerns. They do not place great importance on land. They are older than farmers in the other groups and have been active in farming much longer. They are therefore much closer to the actual time of inter-generational change and therefore have a much higher level of certainty of family continuity. Given their longer time in farming, they are also likely to have the largest asset base and are more likely to have the financial means to expand ensuring that the new generation has a viable resource base. They have therefore been more likely to buy or lease more land and water to maintain their irrigated area, and in the future expect to continue to buy or lease more water to expand their irrigated area to meet the need of the next generation.

6 Strengths and limitations

The strength of this study is that it has been iterative, using successive methods – both qualitative and quantitative – to focus on the problem of identifying the influences of non-profit-maximising values on farmers' behaviour.

Limitations to this study are that the research was undertaken at a time of change in policies concerning water and water trading. It may have been that the surveys were affected by this policy uncertainty as well as the continuing drought and low cotton prices. These results could well be different with other farmers in another policy context. Despite these limitations, the study found three convincing value constructs that were shown to influence farmers' decisions and underpin three cluster groups of farmers who behave differently according to their values.

The study relied on the development of a new instrument to measure irrigators' non-profit-maximising values. If an instrument had already existed that it was capable of measuring the importance irrigators placed on these values, using it would have been preferable. The development of valid and reliable instruments takes time, patience and knowledge. It is only when reliable and valid instruments are not available to measure a particular construct of interest that we might turn our attention, albeit reluctantly, to instrument development (Pett *et al.*, 2003).

That the instrument needs further refinement is shown by cluster groups, which although useful for explaining the non-profit-maximising influences on farmers' behaviour, were perhaps not as effective as originally expected.

The rudimentary stage of development of this instrument has played a part in reducing the strength of the relationships. However, further development of the instrument is currently under way because the encouraging results provided by this study have shown that it is warranted.

7 Policy implications

The most relevant, and most immediately useful, application of the findings of this study is to provide possible insight into why irrigators might be reluctant participants in the Australian government's water entitlement buy-back scheme (Wong, 2008), and how future purchase offers might be better designed to increase participation rates. There are two main ways that the results of this study could be useful for informing policy design.

The first is to design policies that take into account how irrigators are motivated by each of the value constructs. The importance of the value constructs is that they illustrate decision-making influences coming from the main idea of the value construct. (The value constructs are groupings of like values that are connected, congruent and consistent with, and supportive of, the value constructs' main idea.) However, this focus on one main idea means that, for most irrigators, each value construct is only able to explain a part of what influences their behaviour. As discussed previously, the results show that most irrigators are influenced by each of the three value constructs to a greater or lesser extent. Even though most irrigators are pursuing objectives associated with more than one value construct – what Hatfield-Dodds *et al.* (2007) label as "multiple contested values" – policies and instruments could still be designed to allow irrigators to achieve the objectives aligned with a specific value construct. Whether a specific irrigator will react to the particular policy or instrument will depend on how their decision is influenced by the value construct at that particular time. Using this knowledge, the policy-maker will know that by designing a particular instrument or policy in a specific way the policy will be more effective in influencing those who consider that particular value construct to be important.

The second way to design policies is by visualising an actual irrigator such as described by the cluster groups. The cluster groups are distinct groups of irrigators who are influenced by a range of values. The cluster groups were not formed using the factor scores, but they were formed using the respondents' ratings of the original value statements. The values influencing the cluster groups do not completely replicate the values found in the value constructs but are very consistent with them (see Figure 2).

Value constructs are helpful for understanding the components of the non-profit-maximising influences on farmer's behaviour but are not quite as useful for describing how farmers actually behave. Within the limits imposed by the choice of cluster variables, the cluster groups are better able to explain a more complete range of influences on the individual than the value constructs can. The cluster groups, because they are more representative of reality, are therefore a more appropriate choice for thinking about people. The best approach may be to design policies with the cluster group members in mind and with the awareness that they are irrigators who are influenced by certain combinations of value constructs.

The usefulness of this analysis for policy development is the recognition that each of these three groups may behave differently when presented with the opportunity to sell water to the government, participate in other similar reform programs or adopt a certain technology or management practice. When these different influences on irrigators' behaviour are acknowledged, it becomes possible to consider developing more sympathetic policies and programs tailored to the actual values and attitudes of farmers. Using the sale of water as an example, the next paragraphs describe the different ways in which each of the cluster groups may behave.

The *Investor* can be expected to make decisions based on the financial benefits associated with the sale of their water. They will make these decisions recognising that water plays an important part in achieving their businesses goals of profit (they perceive the *Commerce* value construct to be very important). An approach that the government could employ to ensure greater participation in the water entitlement buy-back scheme from this group would be to ensure that profitable business opportunities can be preserved; at least until the time of their retirement. One of the interviewees from this group illustrated his feelings about how the sale of water fitted into his retirement planning, suggesting, "When I want to retire to the Gold Coast . . . that water licence alone is going to be worth a hell of a lot".

The *Expander* will be less likely than the *Investor* to sell their water as it plays a part in achieving their goals of profit and family succession (they perceive the *Commerce* and the *Succession* value constructs to be important). Water is used by this group to generate a profit that assists them in developing a viable business for family successors. The government could gain greater participation from this group by developing programs and policies that recognise that their motivation comes from a need to satisfy a combination of profit and family goals. An interviewee from this group described how he would not consider selling his water because it was useful for his sons in continuing the business, stating ". . . I don't believe we will ever sell the water licenses . . . I want the boys to carry on with it [the farm]; they've helped build it . . ."

The *Traditional*, who is primarily motivated by family succession (they place a high importance on the *Succession* value construct), need the water to provide a viable business for family successors. For this group, water is tied to the achievement of family goals. If the government wants this group to engage in the water entitlement buy-back scheme, they could consider developing programs and policies that allow them to continue as farmers, and even to further develop the family farm, but not as an irrigation farm. Because they have a feeling for the land (which means that they consider the *Caretaking* value construct to be important), the farming culture, and already have a large component of their farm business based on dryland farming, they could be encouraged to expand that part of their business. Also because some of the members of this group are certain of not having family succession, some could be encouraged to remain on their properties as dryland farmers by paying them to provide ecosystem services (as suggested by Bjornlund, 2004). Illustrating the importance and inter-relationships of land, water and family for the *Traditional*, one interviewee from this group stated that "[n]ow my son's put his name on a piece of dirt; I would never consider selling my water".

Policy-makers could take advantage of these findings by understanding that irrigators are influenced by non-profit-maximising values, recognising the motivations underlying each of the value constructs and developing policies that take into account the different motivations of each of the three groups of farmers. One example of how policies and programs could be designed to more accurately target irrigators was announced by the Australian government on 20 September 2008. The "Small Block Irrigators Exit Grant" is a program that will

> *... provide struggling small irrigators—many of whom have been carrying on farming enterprises that were established by their families several generations ago—with an opportunity to exit the sector while remaining in their family homes.*
>
> *Many of these irrigators live close to rural and regional centres. Staying on the farm and working nearby are real options for them.*
>
> *The objective is to help these small block irrigators remain in their communities while getting out of the business of irrigation, at the same time as providing greater environmental flows to the Murray-Darling system* (Prime Minister of Australia).

The conditions of the $150,000 grant are a maximum farm size of 15 hectares, and also that irrigators sell their water to the government, remove their plantings and their irrigation infrastructure, and not re-enter the irrigation industry for a period of 5 years (Department of the Environment Water Heritage and the Arts). The grant allows applicants to retain their house and land but not to operate an irrigation business. Unlike a previous exit package that was not specifically designed for irrigators (Australian Government), elements of this plan could appeal to those with a strong attachment to land or community (see Table 1) because it does not require them to sell their house and land.

The aim of this chapter has been to identify how the non-profit-maximising values that irrigators consider important such as their concerns for family, profit, land, water and community influence their behaviour.

8 Recommendations for future research

While this study has made encouraging first steps in examining the values related to family, land and profit, and to a lesser degree, community, water and lifestyle and their relationship with famers' behaviour, these values need to be examined in greater detail so that they are effectively understood. Future work is needed to

- validate, extend and test the results of this study,
- investigate the extent to which farmers' non-profit-maximising values actually influence their decision-making behaviour,
- investigate the directionality of the relationships between farmers' non-profit-maximising values and their decision-making behaviour,
- group farmers into typologies depending on the values influencing their decisions, identify how farmer types differ in their decision-making, identify how value constructs vary between industry groups and determine the proportion of all farmers that belong to each group, and
- investigate how these values are formed, how entrenched they are and how (or if) they can be influenced and changed.

Some of these aims could be achieved using a case study approach focused on a small number of cases from each of the types identified by this study. Some of the aims could also be achieved through the use of experimental economics, where different groups of farmers are presented with a range of incentives packages in a controlled environment. The value constructs, the cluster groups and the relationships discussed in this chapter show consistent trends in the data that indicate that the above questions would be beneficially pursued in future research.

9 Conclusion

Because sustainable irrigation requires that the irrigation industry continues to generate reasonably priced agricultural products into the future while not degrading the environment or eroding the viability of their communities; irrigators will be required to make changes to their existing practices. In Australia, these changes have begun, are gathering pace, and will, by necessity, become much more widespread.

These extensive changes will affect how irrigators are able to access and use the water resource that they have often taken for granted. If irrigators are to be able to respond to these changes, and maintain or even increase their existing levels of production, tools need to be in place to allow them to adjust to reduced levels of water access. The government has had an optimistic reliance on water markets to facilitate this adjustment, but that approach has assumed that irrigators behave as profit-maximising actors.

The findings reported in this chapter have shown first that the values farmers consider important enough to influence their behaviour, including those of profit maximisation, can be identified and grouped, and second that farmers can be grouped according to which of these values they consider to be important and that these groups can be useful for the purpose of predicting how farmers might participate in programs of reform aimed at achieving sustainable irrigation.

References

ACIL Tasman. 2008. *Australia's Working Rivers: The Role of Infrastructure and Water Buybacks in Recovering Environmental Flows.* ACIL Tasman Pty Ltd, Melbourne, Australia.

Australian Government. 2008. Exceptional circumstances exit package, Canberra, Australia, http://www.centrelink.gov.au/internet/internet.nsf/filestores/se128_0710/$file/se128_0710en.pdf.

Aqualina, J. 2003. Water Sharing Plan for the Upper and Lower Namoi Groundwater Sources, 2003 Order.

Bjornlund, H. 2004. Half a billion dollars—how about a water bank? *Water,* 31(5): 18–19.

Bjornlund, H. 2006. Increased participation in Australian water markets. In: Lorenzini, G. & Brebbia, C. (Eds.), *Sustainable Irrigation Management, Technologies and Policies.* WIT Press, Southampton, pp. 289–302.

Bjornlund, H. 2008. Markets for water allocations: outcomes and impacts. *Proceedings from the International Conference, Water Down Under,* Adelaide, Australia, April, 2008.

COAG. 1994. Council of Australian Governments' Communiqué, 25 February, 1994.

COAG. 2004. Council of Australian Governments' Communiqué, 25 June, 2004.

Department of the Environment Water Heritage and the Arts. 2008. Murray–Darling Basin small block irrigators exit grant package, Canberra, Australia, available at http://www.environment.gov.au/water/programs/entitlement-purchasing/small-block-irrigators.html.

Hatfield-Dodds, S., Syme, G. & Leitch, A. 2007. Improving Australian water management: the contribution of social values research and community engagement. *Reform: A Journal of National and International Law Reform,* 89: 44–48.

Kuehne, G. 2009. The influence of values and attitudes towards family, land, water, community, lifestyle, and profit on farmers' behaviour. Business, University of South Australia.

Kuehne, G. & Bjornlund, H. 2006. Frustration, confusion and uncertainty: qualitative responses from Namoi Valley irrigators. *Water,* 33(4): 51–55.

Kuehne, G., Bjornlund, H. & Cheers, B. 2007. Custodians, investors and conservers: three types of irrigator on the lower Murray reclaimed irrigation areas. *International Journal of Interdisciplinary Social Sciences,* 2(2): 179–186.

Kuehne, G. & Bjornlund, H. 2008. Exploring non-profit maximising influences on irrigators' behaviour. Presented at Irrigation Australia Conference 2008, Melbourne, Australia.

Kuehne, G., Bjornlund, H. & Cheers, B. 2008. Identifying common traits among Australian irrigators using cluster analysis. *Water Science and Technology,* 58(3): 587–595.

Murray–Darling Basin Ministerial Council. 1995. An audit of water use in the Murray–Darling Basin. Murray Darling Basin Ministerial Council, Canberra, Australia.

Namoi Groundwater Taskforce. 2000. The final report of the Namoi Groundwater Taskforce.

Namoi Groundwater Management Committee. 2001. Draft water sharing plan for Upper and Lower Namoi groundwater sources.

NSW Government. 2000. 'Water Management Act 2000', No. 92.

Pett, M., Lackey, N. & Sullivan, J. 2003. *Making Sense of Factor Analysis: The Use of Factor Analysis for Instrument Development in Health Care Research.* Sage: Thousand Oaks, California.

Pigram, J. 2006. *Australia's Water Resources: From Use to Management*. CSIRO Publishing, Collingwood, Victoria, Australia.

Powell, R., Thompson, D., Chalmers, L., Gabbott, A., Stayner, R. & McNeill, J. 2003. A socio-economic analysis of the impact of the reductions in groundwater allocation in the Namoi Valley—final report. Centre for Agricultural and Regional Economics Pty Ltd, Institute for Rural Futures, UNE, Armidale, NSW, Australia.

Prime Minister of Australia, 2008. New exit grant package to assist small-scale irrigators, Canberra, Australia, available at: http://www.pm.gov.au/media/Release/2008/media_release_0487.cfm.

Waterfind. 2008. *Waterfind Analysis of the Federal Government Buyback*. Waterfind Pty Ltd, Adelaide, Australia.

Wong, P. 2008. Water for the future, Sydney, 2008.

Young, M. & McColl, J. 2007. The unmentionable option: is there a place for an across-the-board purchase? *Droplet*, 8(22): 1–2.

Chapter 7

Towards more sustainable irrigation: Factors influencing allocation and entitlement prices and demand in the Goulburn Murray Irrigation District of Australia

M.P. Shanahan[1], S. Wheeler[1] & H. Bjornlund[1,2]
[1]*Centre for Regulation and Market Analysis, University of South Australia, Adelaide, Australia*
[2]*University of Lethbridge, Lethbridge, Alberta, Canada*

Abstract

For well over a decade, water markets have been used in Australia to reallocate water to more efficient, productive and sustainable irrigators. In the Goulburn Murray Irrigation District, records have been kept since 1992, and trading in both water allocations (seasonal trading) and water entitlements (permanent trading) is permitted. This chapter discusses the published quantitative estimates of farmers' responses to changes in prices, focusing particularly on two periods of drought (2002–03 and post 2006). Changes in the farmers' demand and supply of water during drought provide insights into their behaviour in the future when it is anticipated water will be increasingly scarce. We conclude that we can predict some of the likely effects, and the likely sustainability of farm production, in a future with less available water for farm production, by learning from the responses of farmers to past periods of scarcity.

Keywords: Allocations; Drought; Entitlements; Water markets

1 Introduction

This chapter explores water markets' ability to facilitate sustainable irrigation by examining some of the major influences on allocation and entitlement prices and the demand and supply of water in the water markets in Australia's largest irrigation district, the Goulburn Murray Irrigation District (GMID). Using information from the buying and selling of water entitlements (the long-term right to access water) and water allocations (the right to short-term use of water) since the early 1990s, we examine some of the factors that impact on the demand, supply and price of these goods. The insights gained from studies that cover periods of severe drought (2002–03 and post 2006) and periods of reasonable supply (1993–02) allow us to comment on the likely market impact of long-term reductions in water supply to cope with climate change, and the policies that lead to more sustainable water management and irrigation. These insights also allow us to speculate on the likely economic and social consequences of efforts to meet competing objectives: the ability to achieve viable and enhanced ecosystem outcomes while also achieving a sustainable irrigation industry and viable irrigation communities.

2 The study region and water trading background

The GMID is located in Northern Victoria along the River Murray which itself forms the border with New South Wales (Figure 1). Irrigation within the district is mainly supplied by two major sources: the Goulburn and the Murray rivers.

Irrigation began in the district in the 1880s, and in the century that followed, population settlement, agricultural development and irrigation were inextricably

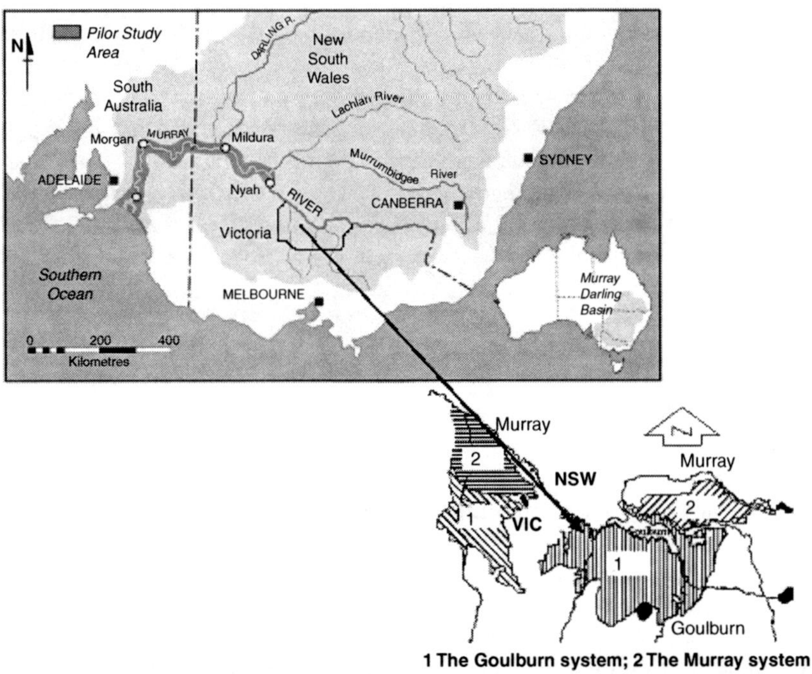

Figure 1: The Goulburn Murray Irrigation District.

linked. Rapid expansion occurred in the district between the world wars, promoted by government policies of closer settlement. Seen as a major region of agricultural production, during the 60 years following the River Murray Agreement of 1915, water regulation was dominated by supply management. Most effort was placed into constructing dams, locks, weirs, pumping stations and distribution networks so as to control, allocate and regulate flows and expand the storage capacity of the river system. Little consideration was given to any natural "limit" imposed by nature until in the 1960s, environmental constraints such as rising water tables and increased salinity began to be viewed as reasons for concern (Langford *et al.*, 1999; Shanahan & Hughes, 2008). Despite state and local efforts to alter governance arrangements and an increase in the number of scientific studies demonstrating a system under strain, response to these issues was slow. It was not until 1992 that a new Murray–Darling Basin Agreement was signed – the first whole of system response since 1915. Around the same time, water management began to change with the partial introduction of water markets. While trade in water was initiated in the early 1980s, early markets were only partially successful. It was not until the real limitation to overall water supplies was appreciated, and farmers began to learn the value of water, that the market became more sophisticated (Langford *et al.*, 1999).

Australia has promoted markets as an integral part of agricultural water management since 1994, although several states began experimenting with markets slightly earlier. Initially water entitlements were bundled with land ownership. Since 1994, they have been separately tradable, and under the National Water Initiative (NWI), they can now be owned separately from land and their worth "unbundled" from land. Each irrigator prior to the NWI owned a volumetric entitlement, but this is now defined as a share of the consumptive pool of water within the relevant water resource.

In the GMID prior to 1998, the water authorities announced the seasonal allocation as a percentage of the underlying entitlement at the beginning of the season, taking historical inflows into account. In 1998, the authorities announced an opening allocation at the beginning of the season influenced by what was in reservoirs, and then each month through the season, the authority revised the allocation, dependent on weather conditions, current water storage levels and anticipated demand. Hence, for the past 6 years, the opening allocation in the Goulburn system has been 0%, and for 5 years before then, below 55% (Bjornlund & Rossini, 2008). This approach, while properly acknowledging the scarcity of water in the system, has significantly increased farmers need to manage the risk associated with the uncertainty of supply and has been a significant driver behind the increased trading in water allocations (Bjornlund, 2006).

The rationale for water markets is simple. Farmers who are prepared to pay can purchase water in the market and produce a crop, while those who are prepared to sell can earn income. In this way, water will be reallocated from areas where it is used to produce low-value crops (e.g. cereals and some grazing and mixed farming) to high-value crops (dairy, fruit and wine) and from less efficient to more efficient irrigators. Trade in water allows some farmers to earn an income when using their own water for irrigation is unviable and, in the longer term, provides them with additional income should they wish to leave the industry (Bjornlund, 2002a). It also allows other irrigators with permanent plantings or dairy herds to buy water to stay in business during drought, assisting both individual irrigators and their communities to overcome severe water shortages. Trading thus helps "soften the blow" caused by water scarcity and partially alleviates the social and economic consequences otherwise caused by a rapid loss in income. It can also facilitate the movement of agricultural water from poor soils (where irrigation has a negative impact on river water quality) to better soils in more suitable locations and so support the process towards more sustainable irrigation.

Under the first trading rules, there were restrictions on trading of water out of certain districts, with some transfers being prohibited entirely and others being limited to 2% of the total entitlement base at the beginning of each season. Under the NWI, the cap has been increased to 4% and is likely to eventually be removed entirely.

Initially, water trade in the GMID in both the markets for water allocations and entitlements was low (Turral *et al.*, 2005; Bjornlund, 2004). Since then, and in contrast to many other countries, markets in allocations have developed to a high level of maturity. The market for entitlements remained more subdued until water scarcity really took hold in 2007 and 2008 (Bjornlund & Rossini, 2008). This is partly because irrigators perceive water entitlements to be an integral part of the value of the farm and partly due to the uncertainty associated with the long-term level of seasonal allocations yielded by these entitlements (Tisdell & Ward, 2003; Bjornlund, 2003a; 2005). Despite this, trade in both forms of water has increased over time and irrigators have increasingly adopted water trading and especially allocation trading as the market, and its support processes have become more sophisticated (Bjornlund & Rossini, 2008; Bjornlund, 2006). For example, in 1998, a weekly water exchange was introduced to ease the administrative pressure on the authority and facilitate faster and cheaper transactions (Bjornlund, 2003b). In 2003, the exchange was extended to cover all of Victoria and the Murray region of New South Wales. These multiple exchanges were gathered under the umbrella of WaterMove (www.watermove.com.au).

Of the many regions covered by WaterMove, this chapter reports on work that concentrates on the largest and most active zone; the Greater Goulburn trading zone within the Northern Victoria regulated region. The exchange facilitates transactions in a number of different water products. This chapter focuses on trade in water allocations, which occur 10–11 months in every year (with the exception of July, generally the wettest month) and water entitlements, which can occur all year round.

WaterMove conducts water exchanges within specified trading zones. In the allocation market, buyers and sellers submit their bids to buy (sell). The bids must include the amount that they are prepared to buy (sell) and the price at which they are willing to buy (sell) the water. The exchanges are held every Thursday and the rules stipulate that the buyers' bids be stacked in descending price order with the highest bidder eligible to buy first. The sellers' bids are stacked in ascending order, with the lowest bidder eligible to sell first. The pool price is set at the level where the maximum volume is traded while no buyer pays more than their bid and no seller receives less than their bid. The pool price is the same for all successful buyers and sellers, with some buyers paying less and some sellers receiving more than their bids (Bjornlund, 2003b). The market for water entitlements has been greatly simplified following the NWI; buyers and sellers agree on a price and then complete the transaction. Until entitlements were unbundled from land ownership under the NWI, the transfer process was cumbersome and one of the main impediments to the slow uptake of entitlement trading (Bjornlund, 2003a).

By July 2004, more than four-out-of-five farm businesses within the GMID had traded water at some time, and in many areas, the figure was higher than nine in ten. During the very dry seasons of 2002–03 and 2003–04, 60% of all farm businesses were active in the water market (Bjornlund, 2006). Water purchased in the allocation market, as a percentage of total water use within the GMID, has increased from approximately 3% during periods of high supply in the mid 1990s, to more than a third by 2006–07. These are clear signs of increased market adoption among water users.

In most years, irrigators within the Goulburn system have received their full allocation (Table 1), with the exception of 2002–03 and 2006–07, 2007–08 and 2008–09 where allocation as at 1 March 2009 was 31% for the Goulburn and 35%

Table 1: Volume traded as percentage of water use – the Goulburn and Murray systems.

	Goulburn system		Murray system	
Season	Allocation (%)[1]	Trade (%)[2]	Allocation (%)[1]	Trade (%)[2]
1995–96	150	7	200	3
1996–97	200	4	200	3
1997–98	120	9	130	13
1998–99	100	13	200	5
1999–00	100	14	200	8
2000–01	100	16	200	2
2001–02	100	18	200	5
2002–03	57	24	129	16
2003–04	100	16	100	18
2004–05	100	18	100	22
2005–06	100	22	144	14
2006–07	29	37	95	20
2007–08	57	29	43	36

[1]Maximum seasonal allocation; [2]Total water trade for season as percentage of total water use.

Source: Authors' calculations. Based on Goulburn–Murray Water's Records.

for the Murray system. During the drought year of 2006–07, opening allocations were 0%, and by December, had only increased to 24% within the Goulburn system. To use water beyond their allocation, irrigators can purchase water allocations from other irrigators. In general, the volume of water traded has been increasing over time.

The annual volume of water allocations traded has increased from less than 50,000 ML (megalitres) in 1989 to almost 350,000 ML by 2006. While the quantity of water traded has steadily increased, prices have been much more influenced by seasonal conditions, such as the drought in 2002–03. Before 2001–02, average monthly pool prices ranged from \$10 to \$100 ML^{-1} (except for 1 month where it reached \$170 ML^{-1}), and increased to \$480 ML^{-1} in 2002–03, and then falling again during the seasons of 2003–04 to 2005–06 (allocation prices, Figure 2). Large variations are also reflected in the weekly pool price and quantity traded. For example, in 2002–03, the weekly pool price varied between \$500 and \$105 ML^{-1}; the maximum quantity traded in a week was over 2500 ML and the minimum under 200 ML. In contrast, in 2004–05, the weekly price ranged between \$103 and \$26 ML^{-1} while the quantity traded each week ranged from over 4000 to only 10 ML. During the drought season of 2006–07, prices reached in excess of \$900 ML^{-1} by December while the accumulated volume traded at that time exceeded any other year and the volume traded in one single exchange was 13,856 ML (Bjornlund & Rossini, 2008).

Since 1993, there has been a mean annual growth in allocation prices of 20.2% and 12.3% in entitlement prices. As Figure 2 shows, there is also clear evidence that the two prices are closely linked, following the same cyclical pat-

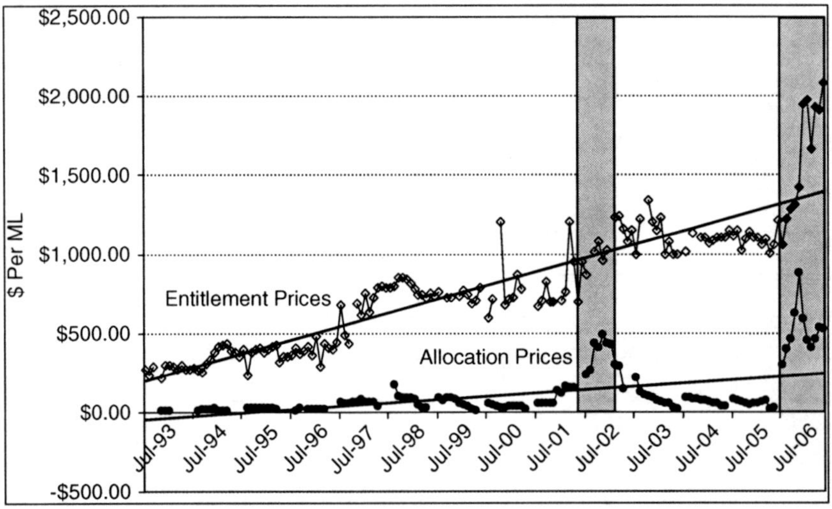

Figure 2: Monthly average prices paid for water allocations and entitlements in the Greater Goulburn from 1993 to 2007 (Periods of drought are shaded).

tern but with allocation prices fluctuating more than twice as much as entitlement prices (Bjornlund & Rossini, 2008; Wheeler *et al.*, 2008c). Although allocation and entitlement prices have increased steadily since water trading began, they have been very variable, and appear strongly influenced by seasonal conditions such as the drought in 2002–03 and 2006–07.

Understanding the factors influencing price determination in water markets is important. While prices are determined by the interaction between supply and demand of water, these are themselves influenced by a range of other policy and market factors. Generally, economic analysis of agricultural prices frequently considers factors such as stocks, imports, production, usage and exports; however, modelling water prices is more complex and there have been few attempts to estimate the factors impacting on prices paid for water in the markets for water allocations and entitlements (Bjornlund & Rossini, 2005; 2007a). Still fewer studies have examined changes to water demand in the market over an extended period that includes drought. A close examination of the factors that impact on water demand in the entitlement and allocation market both in "normal" periods of supply and in drought should provide insight into what may occur as water supplies lessen in the future, either because of climate change or because of government policy to increase the allocation to environmental flows.

3 Influences on allocation and entitlement prices

Previous work by Wheeler *et al.* (2008c) demonstrates that the price of allocations in the GMID is significantly and positively influenced by a yearly trend and the change to government policy introduced in 1998 to facilitate allocation trading. Other potential influences, such as commodity prices, are virtually insignificant (Bjornlund & Rossini, 2005). In any particular month, however, the price of water allocations is negatively and significantly related to the current allocation level. Even allowing for lagged effects (such as the previous month's allocation), current allocation levels are still significant, while the net water deficit (the gap between rainfall and evaporation) also becomes significant. Thus, unsurprisingly, in the short term, the price of water allocations is heavily dependent on the weather (and particularly the size of the net water deficit) and the amount of water available from storages and released by the authorities.

By way of contrast, the price of water entitlements is significantly and positively influenced by several factors including allocation prices (reflecting weather and storage conditions), the month of the year, a yearly trend and the 1998 policy change. Overall, more of the variation in entitlement prices can be explained by the economic models than the variation in allocation prices. The single most significant influence on entitlement prices has been the yearly trend, highlighting the increasing acceptance of trading entitlements, and that over time, investing in such entitlements has proved profitable. The other most significant influenceon entitlement prices is allocation prices (Bjornlund & Rossini, 2007a). As allocations become more expensive, it follows that irrigators are willing to pay higher prices for water entitlements, both to reduce the need to buy future allocations and to attempt to ensure a more secure supply during periods of scarcity.

In summary, over the period 1993–2007, fluctuations in the price of water allocations have been mostly influenced by short-term factors (such as the drought, the current allocation of water received and the net deficit of water) rather than the longer-term prices of water entitlements. Permanent plantings of grapes (and the long-term losses associated with a single drought season) also seem to have influenced entitlement prices in the GMID, but not so much water allocations. Government policy changes had a positive influence on both allocation and entitlement prices, indicating that administrative policy changes can make a significant difference to water prices. Finally, the increasing acceptance of a market approach to water allocation is revealed in long-run increases in entitlement prices as the demand and prices paid for entitlements have increased as irrigators recognise the need to secure long-term water rights.

4 Influences on the elasticity of water demand and supply

4.1 Water allocations

While the price of water can provide information about its relative scarcity, estimates of the elasticity of demand and supply provide information about how farmers respond to changes in prices. In periods of increasing water scarcity, how farmers and their communities respond to increasing prices becomes an important element that affects their overall sustainability.

Elasticity is measured by the percentage change in quantity divided by the percentage change in price. A good that is "inelastic" is one where the quantity does not change a great deal in response to a price change, while an "elastic" response is one where the quantity responds strongly to a change in price. A positive elasticity means that the quantity supplied (or demanded) moves in the same direction as the price change and a negative figure represents a change in quantity in the opposite direction of the price change. For example, a value of 0.5 means that for every 1% increase in price, there is a 0.5% increase in quantity, while a –0.7 means that for every 1% rise in price, there is a 0.7% fall in quantity.

There has been comparatively little work estimating the elasticity of water allocations. Estimates tend to be more elastic if they are based on econometric studies or using mathematical programming and less elastic if derived from models based on field experiments. A recent review of 24 studies on irrigation water in the United States reveals a mean price elasticity of 0.48 and a median of 0.16 (in absolute terms), with larger elasticities in the longer run (Scheierling *et al.,* 2006). They also reveal a large variation in the results, with elasticities ranging from 0.001 to 1.97. Overall, the demand for irrigation water appears to be more elastic than the demand for residential water, a result that is consistent with Australian studies on water price elasticities (Appels *et al.*, 2004).

Over the period 1997–2006, and based on bids registered with WaterMove, Wheeler *et al.* (2008a) found the elasticity of monthly demand in the GMID ranged from –1.32 to –7.82. Using supplier's bids from the same source, the elasticity of monthly demand was estimated to range from 1.13 to 6.88.

Demand for water is closely related to seasonal requirements. The figures reveal that the elasticity of demand for water is relatively low early in the season peaking mid season and declining again towards the end of the season. Such a pattern is consistent with the view that farmers are most keen to purchase water early in the season – newly growing crops require water to survive – but as the growing period unfolds and they are either committed to developing the crop or not, the elasticity of demand increases (Wheeler *et al.*, 2008a).

It is also possible that there are particular price ranges within which farmers (in aggregate) behave differently. Underlying this observation is the idea that as the price of water varies from season to season (being higher or lower on average in some years), so different crops become viable and the farmers then active in the market (i.e. those growing viable crops) may have different responses to price changes. An examination of elasticity by price range reveals that elasticity of demand ranged from –0.90 to –7.03 and the elasticity of supply ranged from 0.11 to 4.09 from the lowest to the highest price range. The elasticity of demand and supply for water depends on the time of the season and market prices. Demand elasticity is higher and varies more during the season than supply elasticity. Demand therefore appears to be dominated by active irrigators responding to fluctuating weather and market conditions, while water sellers and lower value producing irrigators deciding to sell all or a large proportion of their water early in the season dominate the supply side (Wheeler *et al.*, 2008a).

Examining actual water traded against actual prices paid (rather than bid prices) between 1997 and 2006, demand elasticity was estimated between 0.51 in the short run and 0.82 in the long run. Statistically, however, these estimates were insignificant, so that in the aggregate, factors other than just price were found to also influence water demand.

As Figure 2 starkly reveals, drought has a major influence on water prices. Unsurprisingly, in periods of drought, prices rise. Wheeler *et al.* (2008a) also suggest that in periods of drought, the elasticity of demand decreases (the elasticities are lowest in drought years). This is consistent with the response that as water becomes scarcer, those farmers who continue to produce continue to purchase it to maintain production irrespective of price increases. Indeed, depending on the anticipated severity of the drought, and the farmer's financial and personal situation prior to the drought, some continue to purchase water even when it is uneconomic to do so. The decision to continue to purchase water, even when product prices for farm output are below the water price, is consistent with the realisation that where permanent plantings exist, a failure to provide minimal water may mean a permanent cessation of production. This also suggests that the longer a drought continues, the more pressing this issue becomes for farmers with permanent plantings and the more the decision to purchase allocation water becomes, in effect, a decision whether to continue farming. This also has implications for the entitlement price.

4.2 Water entitlements

Whereas work on prices and elasticity of allocation water reveals farmers' short-term responses to variations in water markets, seasonal factors, weather conditions, etc., the entitlement market provides insights on farmers' longer-tem responses to change. Responses in the entitlement market therefore provide insights into the possible direction of change in irrigation and the sustainability of this sector. In Australia, however, this market has been comparatively slow to develop, and any insights from previous research should be considered preliminary.

There has been surprisingly little quantitative work on the factors affecting the price of water entitlements. Analyses of individual transfers in New Mexico found that the most influential factors influencing entitlement prices were urban and industrial activity, the priority of the water right that was traded and the volume of water traded (Colby *et al.*, 1993). An analysis of mean annual prices in Colorado revealed that agricultural commodity prices were significant until urban buyers started to dominate the market after which time other non-farming factors such as the number of housing starts were important (Person & Michelsen, 1994). Within the study region, neither the priority of water rights nor competition from urban and industrial users has been an issue.

Bjornlund (2002b) analysed individual transfers of water entitlements during the early years of water trading to establish the factors influencing individual irrigators' willingness to pay and accept prices in the market for entitlements over two time periods within two different states. He found that as trading was opened up both spatially and between different types of water users and classes of water entitlements, and as market operations became more proficient and intermediaries more active in the market, prices increased and became more consistent across space and entitlement classes. This is consistent with earlier findings in US markets [Gardner, 1985; Brown *et al.*, 1982). He also found that efficient and higher-valued irrigators were willing to pay higher prices when buying, and were capable of demanding higher prices when selling; buyers and sellers in the strongest bargaining position paid lower prices and received higher prices; and older farmers sold at lower prices unless they sold as part of a planned retirement process. Finally, he found that price dispersion in the market in South Australia declined sharply between 1987 and 1996, from approximately 18% initially to 6% in the last year. In comparison, markets in Victoria exhibited a relatively constant dispersion during the first 9 years of trading of around 20%; a rate still within the range experienced for many other commodities (Pratt *et al.*, 1979). These findings suggest that for individual farmers, the market for water entitlements is being impacted by normal economic factors, as would occur in any other market process.

Regression analyses, using mean monthly prices, show very few relationships between commodity prices and prices of water entitlements, with only wine grape prices having a significant positive correlation with price. There also appears to be negative correlations between the price of water entitlements and dairy products which are the outputs from the main high-value industry within the district.

The major factors influencing the price of water entitlements are the price of water in the allocation market, the level of seasonal allocation, wine grape prices and interest rates (Bjornlund & Rossini, 2007b).

There is slightly more research available about the demand for entitlement water. In the GMID, between 1997 and 2003 for example, it was found that rainfall and evaporation, seasonality factors, the price of entitlements, the price of water allocations and the price of milk were statistically significant factors influencing the demand of water entitlements. Although useful, the study did not provide a fully specified model of water demand (Bjornlund & Rossini, 2006).

Wheeler *et al.* (2008b) used a region-level model to estimate the factors correlated with the demand for water entitlements purchased by farmers in the GMID. They found that average demand for water entitlements was significantly and positively influenced by the demand in the previous month and seasonality factors. Surprisingly, other variables such as farm income, rainfall and lagged demand were not significant, although the allocation variable (the percentage of the water entitlement that is available for use at that time) was "almost" statistically significant. The difference between the findings of the 2008 study by Wheeler *et al.* and the 2006 study by Bjornlund and Rossini is most likely caused by the continued and intensifying drought that occurred over the period.

Complicating any analysis of agricultural water markets in Australia over the past 10–15 years has been the impact of drought. This may, however, also provide important insights into long-term sustainability of agriculture in periods of climate change and lessening water. For example, for the period 1997–03, Bjornlund and Rossini (2005) found an expected negative relationship between prices of water entitlements and demand for water entitlements, the work by Wheeler *et al.* (2008b) suggests the price of water entitlements was positively (but weakly) related to water demand. Extending the analysis to include drought years confirms that prices were positively significant, in relation to demand for entitlements. Thus, the beginning of a major (and continuing) drought in 2003, where prices have skyrocketed, appears to have fundamentally changed the relationship between price and demand. Indeed, comparing cycle factors for entitlement and allocation prices confirms that up to 2003 the two factors moved in unison, but since then they have departed (Bjornlund & Rossini, 2007a).

That a major drought should cause a significant shift in the market is not unexpected. The positive sign of the price elasticity, although unusual, signals that buyers would be willing to pay higher prices for larger volumes. It is also consistent with some irrigators "digging in" (and paying what it takes) to maintain their farms in the longer term, while others sell. Increasing price and volumes also helps lower average costs, thus favouring larger and more efficient irrigators. It may thus reflect the shift in farmer production processes desired by policy makers. Farmers who are willing to purchase entitlement water in times of drought are also signalling a long-term commitment to irrigation production – something that is only viable for high-value crops. Comparatively "short-term" price changes are then less of a factor influencing the demand for water as farmers restructure their production process.

Between 1997 and 2007, the market has been dominated by historically low allocation levels and the price of water entitlements does not appear to play a significant role in influencing the demand of water entitlements. The regressions suggest that scarcity is the major reason. What is less understood, however, are the factors associated with individual irrigator decisions. For example, during a drought period risk-averse irrigators, in an effort to secure their production, are likely to be buyers in the market, but identifying the cluster of characteristics associated with risk aversion remains unresearched. During the 2002–03 season, there were reports of a high level of anxiety among irrigators on the day of the weekly exchange, as water was needed "instantly" (Bjornlund, 2003a). High anxiety and high prices during this season saw many people active in the entitlement market. Identifying individual farmer characteristics (such as the age of the farmer, attitude towards leaving the farm, their stress and income levels) may play a more important role in determining trade outcomes than previously thought.

There is evidence of increased market activity following seasons of very high allocation prices and difficulties in obtaining water in the allocation market (Wheeler *et al.*, 2008a). Identifying the characteristics of the farmer likely to adjust their entitlement holdings and in which direction, particularly in periods of water scarcity, remains to be done.

To this point, we are not aware of research that has identified significant coefficients for either the short-run or long-run elasticity of entitlement water in Australia. The strongest factor associated with these elasticities has been the time of year, as water entitlements are made effective on 1 July each year and there is no dispute about who has access to the next season's water allocation on this date (Wheeler *et al.*, 2008b).

Drought increases the stress placed on farmers. We also suggest that the elasticity of demand for entitlement water will decrease (become more inelastic), as prices increase and remaining farmers become more determined to purchase entitlements with less regard to price. An important (and yet little researched) factor likely to be influential to this decision is the response of banks and financiers to individual farmers requests for loans. Without adequate access to finance, the ability of the market to redistribute water permanently and to areas of higher productivity is reduced. However, at the same time, the entry into the market by the Australian government as part of its "Water for the Future Plan" may also decrease the elasticity of demand for water and cause significant increases in water prices. The government anticipates spending A$3.1 billion on purchasing water from farmers to address over-allocation and achieve environmental objectives within the Murray–Darling Basin. This means, based on the trend in sales of water entitlements over the past 5 years, the government will need to purchase 100% of current volume of water traded for the next 14 years to achieve its objective of buying 1500 GL (gigalitres) of water. One consequence of this policy action is that it could potentially prevent the market from achieving the water market reform benefits experienced in the past.

5 Conclusion

From the results, it is clear that in periods of great water scarcity, farmer's response to changes in water prices differs from periods when supply is relatively more plentiful. In periods of reasonable supply, farmers' decisions to buy or sell water allocations are most influenced by the time of the season, the condition of their infrastructure, market prices for their production, etc. As supply decreases under increased water scarcity, the pressure on irrigators to choose between maintaining their production and ceasing production increases. This pushes more farmers into choosing whether to produce in a particular season, and hence also impacts on the supply of water available for purchase. As a drought period continues, however, more farmers are required to choose between maintaining production and *permanently* ceasing production – a factor that influences the entitlement market. As individual farmers make these choices both short and long term, they are influenced by a range of additional factors such as their age, whether they have a succession plan for their farm, income (and savings), the state of the market for their crop and the current policy directions of the government. Many of these factors are comparatively poorly understood – despite their importance in determining whether agricultural production will remain viable in the long term.

To date, the period of concurrence between the existence of drought conditions and actively operating water markets in the GMID has been relatively brief. The data we have to this point include the drought years before 2002–3 and after 2006. Nonetheless, we have some insight into farmers' responses.

Current predictions suggest that there will be a permanent reduction in annual water supply in the whole of the Murray–Darling Basin (including the GMID) due to climate change. Second, there is likely to be a permanent reduction in the amount of water available for farming, as the natural environment's needs are given higher priority and the government enters the water market in a significant way. Both of these factors will cause the price of water, in both the entitlement and the allocation markets, to rise to permanently higher levels. While farmers must adjust to these shifts, it is likely that the price of agricultural output will also rise. For policy makers attempting to facilitate a sustainable level of agricultural output, the critical issue is to ensure that water continues to be transferred to where it is most valuable and that change occurs at a rate that allows farmers time to respond (by modifying their demand for water, altering production, investing in more efficient infrastructure and leaving the sector, etc.). Which farmers have the capacity and ability to respond to increased water scarcity, and how quickly they will react to the market remain largely unanswered questions. Nevertheless, much of the future sustainability of the irrigation sector in the GMID depends on the answers.

The future sustainability of irrigation in Australia depends on achieving a long-term equilibrium between the competing demands of agriculture, the environment and the community. It is clear that over the past decade and a half, water markets have assisted irrigators in the GMID to begin to adjust to water scarcity and to move towards more sustainable irrigation.

Acknowledgments: This chapter is based on several projects funded by the Australian Research Council and six industry partners: Murray-Darling Basin Commission, Department of Natural Resources; Department of Sustainability and Environment, Goulburn-Murray Water, Department of Water, Land and Biodiversity Conservation and UpMarket Software Services. The authors would also like to thank Dr Alec Zuo for his extensive econometric assistance and Dr. Arnab Gupta.

References

Appels D., Douglas, R. & Dwyer, G. 2004. *Responsiveness of demand for irrigation water: a focus on the Southern Murray–Darling Basin.* Staff working paper, Australian Government, Productivity Commission.

Bjornlund, H. 2002a. The socio-economic structure of irrigation communities—water markets and the structural adjustment process. *Rural Society,* 12(2): 123–145.

Bjornlund, H. 2002b. Signs of maturity in Australian water markets. *New Zealand Property Journal,* July: 31–46.

Bjornlund, H. 2003a. Farmer participation in markets for temporary and permanent water in southeastern Australia. *Agricultural Water Management,* 63(1): 57–76.

Bjornlund, H. 2003b. Efficient water market mechanisms to cope with water scarcity. *The International Journal of Water Resource Development,* 19(4): 553–567.

Bjornlund, H. 2004. Formal and informal water markets—drivers of sustainable rural communities? *Water Resources Research,* 40: W09S07.

Bjornlund, H. 2005. Irrigators and the new policy paradigm—an Australian case study. *Water Policy,* 7(6): 581–596.

Bjornlund, H. 2006. Increased participation in Australian water markets. In: Lorenzini, G. & Brebbia, C.A. (Eds.), *Sustainable Irrigation Management, Technologies and Policies.* WIT Press, Southampton, pp. 289–302.

Bjornlund, H. & Rossini P. 2006. An empirical analysis of factors driving outcomes in markets for permanent water—an Australian case study. *Proceedings from the 12th Annual Conference of the Pacific Rim Real Estate Society,* Auckland, New Zealand, available at: http://www.prres.net/

Bjornlund, H. & P. Rossini. 2005. Fundamentals determining prices and activities in the market for water allocations. *The International Journal of Water Resource Development,* 21(2): 355–369.

Bjornlund, H. & Rossini, P. 2007a. Tracing evidence of rational investor behaviour in water markets. *Pacific Rim Property Research Journal,* 13: 344–360.

Bjornlund, H. & Rossini, P. 2007b. Fundamentals determining prices in the market for water entitlements—an Australian case study. *International Journal of Water Resources Development,* 23(3): 537–553.

Bjornlund, H. & Rossini, P. 2008. Are the fundamentals emerging for more sophisticated water market instruments? *Proceedings from the 14th Annual Conference of the Pacific Rim Real Estate Society,* Kuala Lumpur, Malaysia, January, 2008, available at: www.prres.net.

Brown, L., McDonald, B., Tysseling, J. & Dumars, C. 1982. Water reallocation, market proficiency, and conflicting social values. In: Weatherford, G.D., Brown, L., Ingram, H. & Mann, D. (Eds.), *Water and Agriculture in the West U.S.—Conservation, Reallocation, and Markets.* Westview Press, Boulder, pp. 191–256.

Colby, B., Crandall, K. & Bush, D. 1993. Water right transactions—market values and price dispersion. *Water Resources Research*, 29: 1565–1572.

Gardner, B.D. 1985. Institutional impediments to efficient water allocation. *Policy Studies Review*, 5(2): 353–363.

Langford, K.J. Foster, C.L. & Malcolm D.M. 1999. *Toward a financially sustainable irrigation system. Lessons from the State of Victoria, Australia, 1984-1994*. World Bank Technical Paper No 413. The World Bank, Washington, DC.

Person, P. & Michelsen A. 1994. *Determinants and Trends in Water Right Prices: An Econometric Analysis*. Wyoming Water Resources Center, Laramie, Wyoming.

Pratt, J.W., Wise, D.A. & Zeckhauser, R. 1979. Price differences in almost competitive markets. *The Quarterly Journal of Economics*, 93(2): 189–211.

Scheierling, S.M., Loomis, J. & Young, R. 2006. Irrigation water demand: a meta-analysis of price elasticities. *Water Resources Research*, 42: W01411.

Shanahan M.P. & Hughes, S. 2008. A brief history of the supply, demand and sustainable use of water from the Murray–Darling river system. Paper presented to The Asia Pacific Economic History Conference, Melbourne, Australia, February, 2008.

Turral, H.N., Etchells, T., Malano, H., Wijedasa, H. & Taylor P. 2005. Water trading at the margin: the evolution of water markets in the Murray Darling Basin. *Water Resources Research*, 41: W07011.

Tisdell, J., & Ward, J. 2003. Attitudes toward water markets: an Australian case study. *Society and Market Resources*, 16(1): 63–77.

Wheeler, S., Björnlund, H., Shanahan, M.P. & Zuo A. 2008a. Price elasticity of water allocations demand in the Goulburn–Murray Irrigation District. *The Australian Journal of Agricultural and Resource Economics*, 52: 37–55.

Wheeler S., Björnlund H., Shanahan M.P., and Zuo A. 2008b. Price Elasticity and the Demand for Water Entitlements in the Goulburn-Murray Irrigation District of Australia, Water 4th International Conference on Water Resources and Environment Research (Water Down Under) Adelaide, 14–17 April.

Wheeler, S., Bjornlund, H., Shanahan M.P., and A. Zuo. 2008c. *Factors influencing water allocation and entitlement prices in the Greater Goulburn Area of Australia* Sustainable Irrigation Conference 08, Alicante, Spain, 11–13 June.

Chapter 8

Incorporating the environment into the market: The case of water trusts and environmental water transfers in the western United States

Ereney Hadjigeorgalis
United States Department of Agriculture, Washington, D.C., USA

Abstract

As population growth and urbanization accelerate, it is becoming increasingly difficult to meet competing demands for water resources without detrimental impacts on the environment. Nonetheless, a healthy aquatic ecosystem is essential to the sustainability of irrigation and food production in the coming decades. One solution that holds promise for environmental water flows is to incorporate the environment into the market. This has been accomplished to a certain extent in the western United States where federal, state, and nonprofit entities have engaged in environmental water transfers to augment stream flows. This chapter discusses the evolution of the market for environmental water in the western United States and the institutional structures that have both facilitated and obstructed efforts to restore and enhance stream flows. It reviews the institutional and legal basis of water management in the western United States as espoused in the doctrine of prior appropriation, the beneficial use clause, and the provisions of the Endangered Species Act with respect to water flows. It then gives a description of federal provision of water for in-stream flows as well as the various nonprofit water trusts that have arisen in the Pacific Northwest. This chapter concludes with a look to the future of private efforts to fill a niche in environmental resource management in the western United States.

Keywords: Endangered species act; In-stream flows; Water markets; Water trusts

1 Introduction

Water scarcity and water quality have risen to the forefront of environmental issues. All in all, one-third of the world's population lives in basins where water use exceeds sustainable levels, which translates into less water for irrigation, municipal drinking water, industrial production, hydropower, and, lest we forget, sustenance of fish and wildlife populations and recreational uses (UNHDR, 2006).

As population growth accelerates and urbanization increases, this scenario will intensify, placing increasing pressures on agriculture and the environment. Pressures on agriculture will increase because food production is highly water intensive, while agriculture controls the largest proportion of water resources in most basins – 70–85% on average. Pressure on the environment will increase for a very different reason – the environment typically stands in the way of human production and consumption and is the first sector to have its water degraded in the stiff competition for scarce resources.

The secondary status traditionally attributed to environmental water uses is particularly stark in the western United States where water property rights regimes have tended to view only diversionary uses of water as beneficial. Built on a history of large, government–subsidized irrigation and hydropower projects, the West largely ignored environmental flows during the first half of the 20th century as the federal government embarked on an era of rapid dam building. Beginning in 1902, the Bureau of Reclamation alone constructed over 600 dams in 17 Western states, as well as an extensive network of water distribution infrastructure (BOR 2). The construction of these large, subsidized irrigation projects however soon proved economically untenable. While the federal government spent close to $23 billion on water supply projects, only $2 billion was ever recovered from irrigators. By the time all was said and done, the federal irrigation subsidy had reached an effective rate of 85–90% (Neuman, 2004).

By the 1970s, it became clear that the federal government's focus on supply-side management of water resources would be insufficient to manage the mounting water crisis, and there was a paradigm shift towards demand-side approaches based on prices and the concept of water markets. The first operative water markets appeared in the early 1980s, and since that time, they have proliferated across the western United States, Australia, and several developing countries such as Chile, South Africa, China, and South Asia (Hadjigeorgalis, 2009a). In Australia and the United States, these early markets focused on obtaining water supplies from agriculture for growing cities, whereas in developing countries, farmer-to-farmer trades were emphasized (Hadjigeorgalis, 2009b). In neither developed nor developing countries, however, was much attention paid to environmental or in-stream flows.

As Bjornlund so aptly points out in the introduction to this book, however, without healthy ecosystems, water resources cannot continue to meet a constant or growing level of needs into the future. The challenge, as always, is to maintain production output with access to less water while preventing negative environmental impacts and allowing irrigation communities to continue to be viable under a sustainable water use regime.

Institutional structures can clearly have a significant role in facilitating and encouraging this process. For example, incorporating the environment into the market could go a long way towards achieving sustainable irrigation while protecting and enhancing aquatic ecosystems that support native fish populations and important fisheries industries.

In both the United States and Australia, the past 15 years have seen an increased emphasis on using the market to buy water to achieve environmental purposes. This chapter discusses the evolution of the market for environmental water in the western United States and the institutional structures that have both facilitated and obstructed efforts to restore and enhance aquatic environments that serve a multiplicity of uses. It begins by reviewing the institutional and legal basis of water management in the western United States as espoused in the doctrine of prior appropriation, the beneficial use clause, and the provisions of the Endangered Species Act (ESA) with respect to water flows. It then gives a description of the environmental water market for in-stream flows that consist of federal provision of water for in-stream flows as well as the various nonprofit and quasi-governmental water trusts that have arisen, particularly in the Pacific Northwest. This chapter concludes with a look to the future of private efforts to fill a niche in environmental resource management in the western United States.

2 Institutional framework for environmental water transfers

2.1 Prior appropriations doctrine and beneficial use

In most Western states, two concepts define Western water law: the prior appropriations doctrine and beneficial use. The prior appropriations doctrine, which is often referred to as "first in time, first in right," is a property rights system based on the concept of seniority. Each water right is assigned a priority date indicating when the right was first appropriated. The priority date defines the seniority of the water right, and the doctrine assures that those with the claim to the oldest water rights receive their allotments first in times of shortage. Beneficial use governs the end use of the water, which has traditionally been defined to include only consumptive or diversionary uses such as irrigation, municipal demand, or hydropower production. A water right that has not been diverted for a specified period of time (usually 4–5 years) is subject to forfeiture.

Both beneficial use and the prior appropriations doctrine present problems for the maintenance of in-stream flows. To protect in-stream flows, an additional legislation was enacted in most Western states that redefined beneficial use to include in-stream uses, such as for fisheries or recreation. This legislation, in and of itself, however, was insufficient in prior appropriations regimes on over-appropriated rivers; any new in-stream water rights created would be the most junior on the river impeding their fulfillment in droughts when they were most needed and thus defeating the purpose of creating the in-stream right in the first place (Neuman, 2004). The critical institutional change needed was to permit the transfer of senior appropriative water rights to in-stream, nondiversionary uses. Such legislation has been introduced in varying forms across the western United States.

2.2 Endangered Species Act

While the prior appropriations act and the beneficial use clause have presented institutional obstacles to protecting in-stream flows, the ESA has been a strong, enabling piece of legislation for such protection. The ESA was passed by Congress in 1973 under the Nixon administration to preserve biodiversity by preventing species extinction. Many endangered or threatened species are aquatic; hence, the ESA has been instrumental in protecting in-stream flows (Gillian & Brown, 1997).

The ESA of 1973 gives the U.S. Fish and Wildlife Service and the National Marine Fisheries Service the power to recover and conserve all forms of plants and animals found to be threatened or in danger of extinction. As such, the ESA is one of the broadest and most powerful environmental laws of the United States. It has also been controversial and the source of intense public debate involving private property rights, state primacy, and natural resource protection (Utah Division of Water Resources, 2001).

The ESA places a significant burden of responsibility for species protection on the shoulders of the federal government by requiring that federal agencies undertake no action that will harm endangered species. Harm is broadly defined under the ESA as any action that kills or injures wildlife by significantly impairing essential behavioral patterns, including breeding, feeding, and sheltering. In subsequent court cases, the Supreme Court concluded that this protection takes precedence over even the primary missions of federal agencies (Gillian & Brown, 1997).

The U.S. Fish and Wildlife Service and the National Marine Fisheries Service have the authority to ensure that other federal agencies comply with the ESA in their operations. Some of the most important federal agencies that they oversee in this respect are those charged with dam operations and hydropower production because these can reduce in-stream flows and impede fish migrations, particularly for salmon and other anadromous fish species. The federal agencies responsible for dam operation and hydropower production include the BOR (Department of the Interior), the U.S. Army Corps of Engineers (USACE), and the Bonneville Power Administration (BPA) (Department of Energy).

To meet the obligations of the ESA to maintain in-stream flows for fish habitat, this troika of federal agencies purchases water from irrigators, usually through state water banks. For example, in Idaho, the BOR is required, via the ESA, to supply 427,000 acre-feet of water per year to the Snake River to improve in-stream flows for salmon (Scarborough & Lund, 2007). In 2000, the BPA was required to restore stream flows for endangered species to help offset the impact of dams constructed for hydroelectric power generation. The BPA accomplished this by outsourcing improvements in in-stream flows to a number of groups who use markets to supply water. Outsourcing allowed the BPA to use a market process to come up with the water it was forced to provide by court order at the least cost and disruption (Scarborough & Lund, 2007).

When all else fails, the ESA can be used to curtail irrigation in basins where low in-stream flows threaten aquatic species. In 2001, a federal court ordered irrigation water to farmers in the Klamath River Basin cut off (Meiners & Kosnick, 2003) in an attempt to meet the needs of two fish species listed under the ESA (Scarborough & Lund, 2007).

3 The environmental water market for in-stream flows

Participants in the environmental market for in-stream flows include federal and state governments as well as nonprofit and quasi-governmental agencies and irrigators. Markets for in-stream flows extend over six hydrologic regions (Rio Grande, Upper Colorado, Lower Colorado, Great Basin, Pacific Northwest, and California) that include 12 Western states (Colorado, New Mexico, Texas, Arizona, Utah, Wyoming, California, Nevada, Oregon, Washington, Idaho, and Montana).

By far the dominant buyers in the environmental water market are federal and state agencies. To date, 95% of all water for in-stream flows is acquired and allocated by federal and state agencies, with federal expenditures accounting for 60% of total expenditures (see Figure 1) (Neuman, 2004;

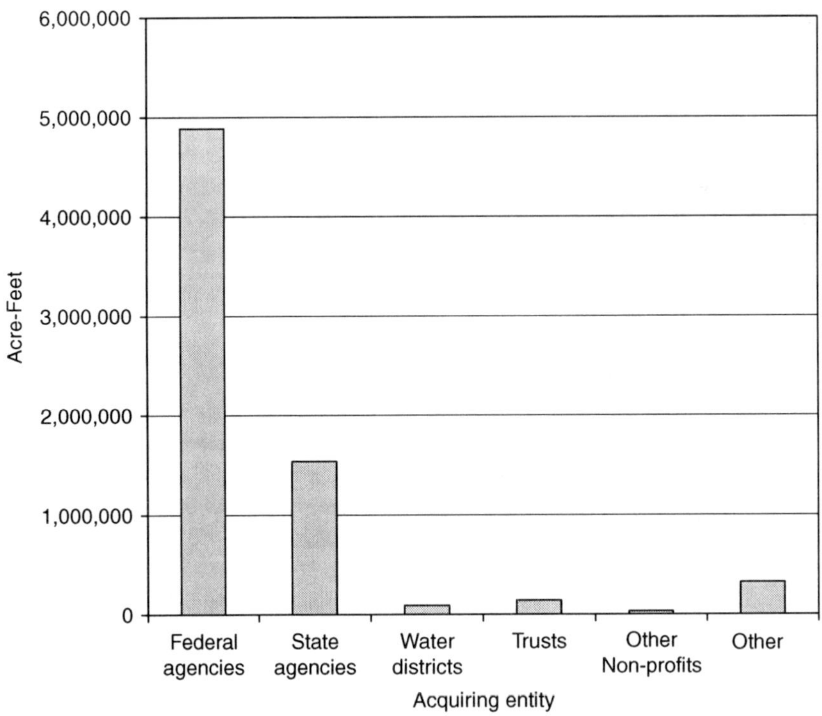

Figure 1: Acquisitions of water for in-stream flows by acquiring entity, 1991–2006. *Source*: Water Strategist.

Scarborough & Lund, 2007). These federal agencies concentrate on procuring large amounts of water for major rivers. From 1998 to 2007, nearly 6 million acre-feet of water had been acquired for in-stream use, almost 2.5 times the amount acquired from 1990 to 1997. In Oregon alone, federal expenditures for in-stream flows exceeded $15 million, restoring nearly 230,000 acre-feet to streams in a series of short-team leases (Scarborough & Lund, 2007). Most of the federal government's acquisition responsibilities are delegated to the BOR and BPA, and much of the activity in in-stream acquisitions is driven by efforts to restore flows for endangered species (Scarborough & Lund, 2007).

The remaining 5% of environmental water market activity is provided through nonprofit water trusts, water districts, and other nonprofit and quasi-governmental agencies. In contrast to federal agency practices, these entities tend to focus on smaller acquisitions of senior water rights on secondary streams that support significant fisheries. Although the water rights that are purchased to maintain in-stream flows are small, the flow is big enough and *senior* enough to assure stream flow in late summer, providing a late-summer rearing habitat for fish and improving water quality (Neuman, 2004).

3.1 Federal and state provision of in-stream flows

In many Western states, only state agencies may acquire and hold in-stream water rights, creating a barrier to participation in environmental water markets by non-profit entities. In these states, federal and state acquisitions are the only source of environmental water market activity. These states include Idaho, Utah, and Wyoming. In other states, private entities may acquire in-stream water rights, but ownership of these rights remains with state agencies – that is, only state agencies can hold in-stream water rights. These include all other Western states that participate in environmental water markets.

3.1.1 Idaho

Only state or federal agencies may acquire and hold in-stream water rights in Idaho. Idaho depends on its water banking statutes to provide for in-stream flows through leases acquired by the Idaho Department of Water Resources or the BOR. The BOR often leases water from the Idaho's water banking system to provide for in-stream flows. From 1991 to 2006, Idaho acquired 2.4 million acre-feet of water primarily through temporary leases – second only to California in total water acquired for environmental flows (see Figure 2). More than 85% of this quantity was acquired by the BOR to meet ESA requirements on the Snake River (Water Strategist).

3.1.2 Utah

Similar to Idaho, only state agencies in Utah have the authority to purchase and hold in-stream water rights. Only the Division of Wildlife Resources and the Division of Parks and Recreation can procure in-stream water rights in Utah. The

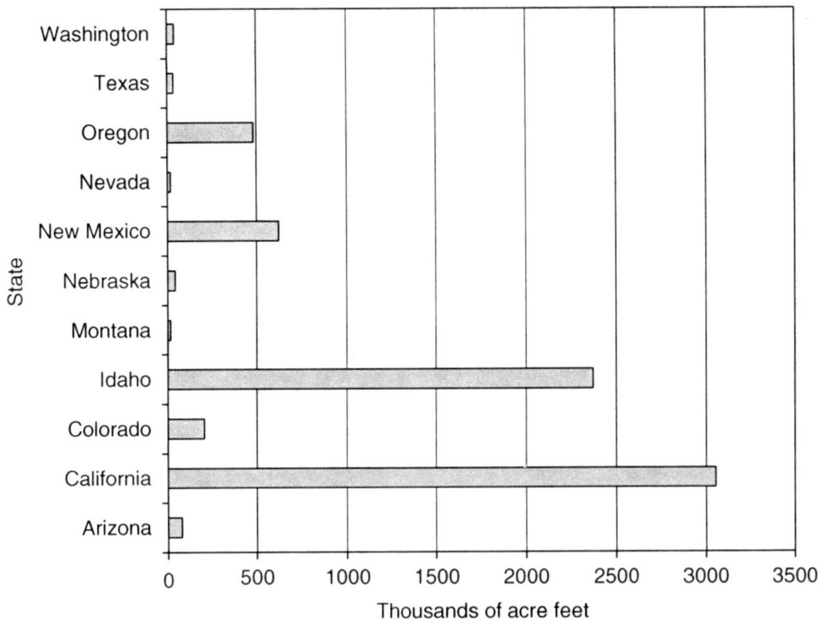

Figure 2: In-stream flow acquisitions by state, 1991–2006.
Source: Water Strategist.

Utah Code allows these two state agencies to file changes on perfected water rights to provide in-stream flows in designated reaches of streams (Utah State Water Plan, 2001). These flows may be acquired for preservation and enhancement of fisheries, the natural stream environment, or public recreation. Because these agencies have limited funding, acquisition of such water rights is dependent on legislative appropriations and a willing seller, unless the water right is previously owned by the agency or is gifted or deeded to it (Utah State Water Plan, 2001). The nonprofit organization Trout Unlimited has been pushing for legislation to allow private entities to purchase water rights for in-stream flows (Scarborough & Lund, 2007).

3.1.3 Wyoming

In Wyoming, in-stream flow rights can only be held by the state, and there are no provisions allowing any private entity to lease or buy water rights for in-stream flows. In contrast to other states, however, Wyoming statutes also do not allow the state to lease or buy water rights for in-stream flows, which limits the state's acquisitions to gifts or voluntary transfers from private water rights holders. To date, no in-stream flow transactions have occurred in Wyoming, and as of 2007, no water right holder had been willing to donate their rights to the state (Scarborough & Lund, 2007).

3.1.4 Colorado

Although the Colorado Water Trust was formed in 2001 to support and promote voluntary efforts to protect and restore the state's stream flows, the greater impetus in Colorado toward the protection of in-stream flows is the Colorado Water Conservation Board, which is the only state entity that can hold in-stream water rights by law (Colorado Water Trust). In 2008, the Colorado Water Conservation Board received a major impetus from the state when the state authorized $1.5 million annually for the purchase of rights for in-stream flows (Smith, 2008).

3.1.5 Southwestern states

In the Southwest, legislation has been unclear with respect to who may obtain and hold in-stream water rights. In Arizona, water law states that water rights may be "transferred for use for ... wildlife purposes, including fish," but no in-stream flow transfers have been attempted. In New Mexico, there is no legislated in-stream flow program and in-stream flow is not recognized as a beneficial use. In 1998, however, the New Mexico Attorney General issued a legal opinion concluding that the transfer of a consumptive water right to an in-stream flow right is allowable under state law. The legal opinion determined that in-stream uses such as recreation and fish and wildlife habitat are beneficial uses, and that transfers of existing water rights to in-stream flows are not expressly prohibited (BLM).

As a result of this unclear legislation, federal and state agencies in the Southwest have been the primary acquirers of water for in-stream flows and much of this activity has been to meet interstate or ESA obligations. In 1994, Arizona's Water Protection Fund was established to protect and restore riparian habitat, including funding for improving in-stream flows. However, almost all of the funding has been directed toward habit restoration efforts rather than acquisitions to restore in-stream flows (Scarborough & Lund, 2007). In New Mexico, the BOR and state agencies such as the Interstate Stream Commission have spent nearly $25 million purchasing land and water rights to improve stream flows for endangered species and reduce interstate dewatering. A law passed in 2005 created a state Strategic Water Reserve that will allow the state to purchase, lease, or accept donated water rights to help endangered species (Scarborough & Lund, 2007). In New Mexico, nearly 250,000 acre-feet of water was put in-stream for fisheries through leases by the BOR and conservation efforts by the Pecos Valley Artesian Conservancy District.

3.1.6 California

More funds have been spent in California to restore in-stream flows than in any other Western state (see Figure 3). Between 1998 and 2005, state and federal agencies spent approximately $156 million to acquire 1.5 million acre-feet of water for in-stream use. Much of this water was transferred to the state's Environmental Water Account, a state agency charged with maintaining in-stream flows to improve spawning habitat for native fish (Scarborough & Lund, 2007). Other market activity has been primarily limited to short-term lease agreements funded by state and federal agencies. Nevertheless, acquisitions for

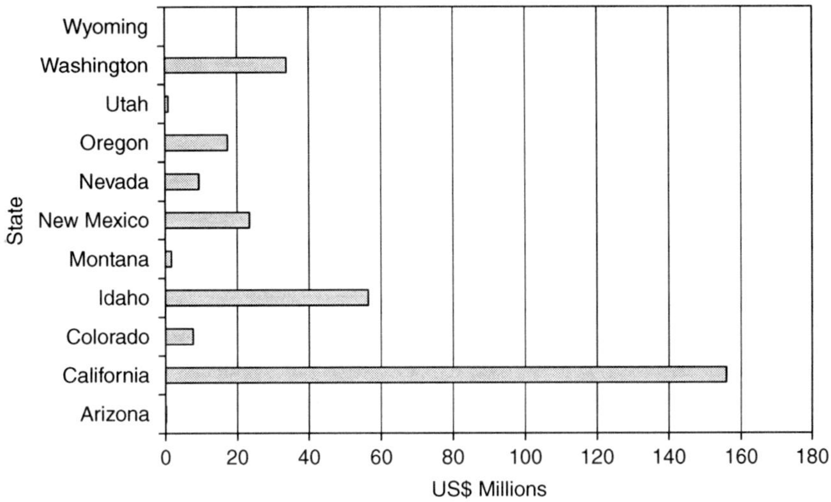

Figure 3: In-stream flow acquisitions in western states, total expenditures, 1998–2005. *Source*: Scarborough, B. & Lund, H. 2007. *Saving our Streams: A Practical Guide, Property and Environment Research Center (PERC)*. PERC, Bozeman, MT.

in-stream flows have been significant over the past 15 years (Scarborough & Lund, 2007).

3.1.7 Nevada

In Nevada, any person may appropriate water for beneficial use, and in-stream flows are recognized as beneficial under state statutes (BLM). Water market activity for environmental flows in Nevada, however, has been largely limited to wetland restoration rather than protection and maintenance of in-stream flows. Two exceptions are purchases of water rights by Washoe County and the cities of Reno and Sparks for transfer to in-stream uses as well as purchases of water rights and land to comply with a 1996 lawsuit settlement in favor of the Pyramid Lake Paiute Tribe against the United States and the cities of Reno and Sparks, NV (Scarborough & Lund, 2007; Nevada Division of Water Resources, 1999).

3.2 Water trusts

From the mid to late 1990s, a complementary movement to the federal provision of water for in-stream flows began to emerge, primarily concentrated in the states of the Pacific Northwest. Known as water trusts, these groups were part of a larger trend of delegation of federal authority to state and local levels, as well as private organizations (King, 2004). They were born out of, and based on, the conservationist land trust model as a derivative of the land trust model espoused by such organizations as the Nature Conservancy and the Sierra Club. Between 1998 and 2005, private organizations were active participants in acquisitions of in-stream flows in only four Western states: Oregon, Washington, Montana, and Nevada.

The rise of water trusts as a complement to federal provision of water for in-stream flows paralleled the listing of several important fish species as endangered. In the late 1990s and throughout the 1990s, population growth and rapid urbanization led to overdiversions of river flows that degraded spawning habitat for many riverine and anadromous fish species and affected water quality. In the Pacific Northwest, several species were listed as endangered or threatened and miles of streams were classified as water quality limited, not only creating environmental concerns but also impacting the economic vitality of northwestern fishery industries.

Despite federal mandates to maintain stream flows under the ESA, the list of endangered and threatened species not only was maintained but also continued to grow. As of early 2009, 26 fish species were listed as endangered or threatened in California, 14 in Oregon, and 5 each in Washington and Idaho. Eighteen animal species are currently listed as threatened or endangered in Utah, of which 8 are fish whose numbers have reached critical levels due to altered flow patterns, water temperatures, and man-introduced predatory species (Utah Division of Water Resources, 2001; FWS). In Nevada, of 104 native fish species, 11 are now extinct or extirpated, while 23 are listed as threatened or endangered (Nevada Division of Water Resources, 1999; FWS).

The most important water trusts include the Oregon Water Trust, the Washington Water Trust, the Montana Water Trust, and the Columbia Basin Water Transactions Program.

3.2.1 Oregon Water Trust

The Oregon Water Trust (OWT) is the oldest water trust in the western United States, and has served as a prototype for other water trusts. The OWT, established in 1993, represents a cooperative alliance between state and local leaders (King, 2004). It is the first private organization in the nation dedicated to acquiring water rights to enhance in-stream flows (OWT). As of 2008, the OWT portfolio had grown from 2 leases totaling 1.4 cubic feet per second to 97 projects amounting to more than 160 cubic feet per second of protected in-stream flows. The OWT works primarily on a preventative basis to restore water for fish habitat before a crisis exists (Neuman, 2004). They concentrate their efforts on smaller streams and tributaries that support significant fisheries, particularly salmonids, and where small increases in stream flow can have significant impacts (OWT).

Two important pieces of legislation enabled the development of the OWT: the In-stream Water Rights Act and the Conserved Water Program, both passed in 1987. The In-stream Water Rights Act of 1987 modified the traditional beneficial use clause of the prior appropriation doctrine to include in-stream uses (Neuman, 2004; King, 2004). The key clause that made this legislation successful where prior legislation had not been was that it allowed for consumptive water rights to be converted to in-stream rights with the priority date of the original right. This gave in-stream rights senior priority status in many cases and allowed such rights to guarantee stream flows in dry seasons on over-appropriated streams. The 1987 law provided the institutional backdrop against which a water market in environmental flows could emerge (Neuman, 2004).

The 1987 Conserved Water Program established the appropriate incentives for water conservation and set up a mechanism for the transfer of a portion of conserved water to in-stream uses. Prior to the passage of the Conserved Water Program, water users who invested in water-conserving technologies had no claim to the water saved by such technologies, which flowed freely in the stream and became available to users with lower priority dates. The Conserved Water Program amended state water laws to allow water users who invested in water conservation measures or technology to retain up to 75% of their conserved water, with the remaining 25% transferred to in-stream flows.

Despite the enabling legislation and ample funding, it proved difficult for the OWT to acquire water for in-stream flows on its founding. To counteract this, the trust began to engage in a number of innovative market devices to create transactions that work in varying circumstances. Examples of these include water conservation projects, split season leasing, source switching, point of diversion change, rotational pooling arrangements, and modified land management (see Table 1).

The OWT spawned the establishment of other water trusts throughout the western United States in Washington, Montana, Colorado, Texas, the Great Basin Area, and New Mexico. In addition, the OWT has actively participated in the development of the Columbia Basin Water Transactions Programs that span Oregon, Washington, Idaho, and Montana.

3.2.2 Washington Water Trust

The state of Washington administers a Water Acquisition Program to aid salmon recovery efforts by contributing to stream flow restoration through the acquisition of water rights on a permanent or temporary basis. The program is intended to be a voluntary, incentive-based initiative, offering monetary compensation to water right holders who are willing to revert all or a portion of their water rights back to the state to increase in-stream flows (Lovrich *et al.*, 2004).

A key partner in the Water Acquisition Program is the Washington Water Trust (WWT). The WWT is a private, nonprofit organization that was established in 1998 to restore in-stream flows in Washington's rivers and streams. The WWT works to benefit water quality, fisheries, and recreation by acquiring existing water rights from willing sellers through purchase, lease, or gift and conferring these rights to the Washington Trust Water Rights Program. As a nonprofit organization, WWT receives funding from various sources, including, but not limited to, the BPA's Columbia Basin Water Transaction Program administered by National Fish and Wildlife Foundation, the Water Acquisition Program under the Department of Ecology, various mitigation funds, private foundations, and private donations (Lovrich *et al.*, 2004).

Table1: Market arrangements used by water trusts.

Purchase	This permanently transfers all or a portion of the water right to an in-stream water right that is held by the state.
Lease	A lease is a temporary acquisition of water rights for in-stream use. In practice, leases have been arranged for periods as short as 1 year and as long as 20 years or more. Most water acquired for in-stream flows is acquired through leases.
Dry year lease	This type of lease allows the farmer to irrigate during wet years while leaving the water in-stream during dry years only.
Split season leasing	Irrigators use water during the first half of the season and lease their rights to in-stream uses during the second half of the season, which is usually more critical for fish habitat.
Water conservation projects	The trust assists irrigators in the installation of more efficient irrigation systems, and in return, the irrigator transfers some or all of the conserved water to in-stream uses.
Source switching	An irrigator switches his water sources from surface water to groundwater or stored water and the surface water right is transferred or leased to in-stream uses.
Point of diversion change	Irrigators agree to withdraw water from a different location on the stream, leaving resources in-stream along the driest and most critical stretches for fish habitat.
Rotational pooling arrangements	Irrigators coordinate with neighbors and take turns leasing water rights.
Modified land management	Irrigators fallow land and/or switch to crops that use less water. Water freed up is leased or sold for in-stream uses.
Nongeneration agreement	Utility agrees not to use their water to generate electricity in exchange for payment for lost revenue.
Replacement feed	Trust provides irrigator with replacement feed in lieu of irrigating pasture for livestock production.

In-stream water uses have been given legal standing as beneficial under Washington Water Law. Therefore, any water rights transferred to the Trust Water Rights Program are protected by the state. Senior trust water rights must be met in-stream prior to the fulfillment of junior uses out-of-stream. Although any entity may acquire water for in-stream flow purposes, all in-stream water rights must be held by the state in the Trust Water Rights Program. The trust water held by the state retains the seniority of the original right and is not subject to relinquishment while in trust status (Lovrich *et al.*, 2004).

3.2.3 Montana Water Trust

In Montana, in-stream uses of water have been recognized as beneficial since the 1973 Montana Water Use Act. However, it was not until 1995 that environmental and agricultural groups formed a partnership to pass legislation that allowed individuals or private organizations to lease water rights for in-stream uses. This legislation enabled the founding of the Montana Water Trust (MWT) in 2001 (Montana Water Trust).

Unlike other states with water trusts, where water may be acquired through donation, lease, or purchase, Montana only allows the leasing of water rights for in-stream uses, prohibiting any permanent transfers. In addition, lease terms are limited to 10 years with the possibility of a one-time 10-year extension. The exception to this requirement is leases of conserved water generated from irrigation efficiency improvements. These leases may be held for the expected life of the irrigation improvement project, or a maximum of 30 years.

Another impediment to the work of the MWT is that they do not have legal standing to enforce the rights that they lease from irrigators for in-stream use. This authority rests only with the owner of the water right, making the leasing of in-stream water rights a more precarious form of in-stream flow protection than in other states.

3.2.4 Columbia Basin Water Transactions Program

In 2002, the BPA created the Columbia Basin Water Transactions Program (CBWTP) to address declining stream flows and habitat loss in parts of the Northwest. While the CBWTP not a water trust itself, it provides funding to organizations in Washington, Oregon, Idaho, and western Montana to acquire water for in-stream flows through purchases or leases. The organizations, in turn, contract with local irrigation districts and landowners to acquire the water (Scarborough & Lund, 2007). The CBWTP has provided funding to the Oregon, Washington, and Montana water trusts as well as the Idaho Department of Water Resources. Since 2002, they have restored more than 1000 cubic feet per second to streams in these states (CBWTP).

4 Conclusion

As agriculture continues to be pressed as a source of water supplies from competing sectors, environmental water transfers will play an increasingly important role in maintaining healthy ecosystems that will be able to sustain food production for future generations. While the rise of nonprofit entities in this arena represents an important environmental movement, it is likely that the bulk of these efforts in the western United States will continue to be driven by federal and state agencies.

The catalyst for the rise of environmental water transfers in the western United States was the ESA, which propelled federal agencies into the role of guardian of in-stream flows. Institutional changes will continue to be the most important component in the coming decades in further facilitating environmental water market transfers to protect in-stream flows. The extent to which nonprofit entities, such as water trusts, will proliferate and flourish will be governed by efforts to revise legislation that codifies in-stream flows as beneficial use. In conjunction with such legislative changes, water markets also need to be liberalized to allow private entities to both purchase and hold in-stream water rights. As was seen in the case of Wyoming, when governments depend on donations of water rights as the basis of their in-stream flow protection, such protection will necessarily fall short. In contrast, when private groups are empowered as they are in Oregon, water trusts can be a critical addition to the protection of in-stream flows in smaller rivers where federal and state efforts are not cost-effective.

References

Colorado Water Trust. www.coloradowatertrust.org. Accessed 27 August 2009.

Columbia Basin Water Transactions Program. www.cbwtp.org. Accessed 27 August 2009.

Gillilan, D.M. & Brown, T.C. 1997. *Instream Flow Protection: Seeking a Balance in Western Water Use*. Island Press, Washington, DC.

Hadjigeorgalis, E. 2009a. A place for water markets: performance and challenges. *Review of Agricultural Economics*, 31(1): 50–67.

Hadjigeorgalis, E. 2009b. The distributional impact of water markets on small farmers: is there a safety net? *Water Resources Research*, 44, doi:10.1029/2007WR006527.

King, M.A. 2004. Getting our feet wet: an introduction to water trusts. *Harvard Environmental Law Review*, 28(2): 495–534.

Lovrich, N.P., Siemann, D., J.B., Bireley, R.M., Gaffney, M.J., Kent, C. & Huckabay, J. 2004. Of Water and Trust: A Review of the Washington Water Acquisition Program. Policy Consensus Center of the University of Washington and Washington State University, available at: http://www.ecy.wa.gov/programs/wr/instream-flows/Images/pdfs/Water%20and%20Trust%20Report.pdf.

Meiners, R.E. & Kosnik, L. *Restoring Harmony in the Klamath Basin, Property and Environment Research Center (PERC) Policy Series PS-27*. PERC, Bozeman, MT.

Montana Water Trust. www.montanawatertrust.org. Accessed 27 August 2009.

Neuman, J. 2004. The Good, the bad, and the ugly: the first ten years of the Oregon Water Trust. *Nebraska Law Review*, 83: 432–484.

Nevada Division of Water Resources 1999. Nevada State Water Plan, 1999.

Oregon Water Trust. www.owt.org. Accessed 27 August 2009.

Scarborough, B. & Lund, H. 2007. *Saving our Streams: A Practical Guide, Property and Environment Research Center (PERC)*. PERC, Bozeman, MT.

Smith, J. 2008. Colorado to spend $1.5 million yearly to protect streams. Rocky Mountain News, September 15, 2008, available at: http://www.rockymountainnews.com/news/2008/sep/15/colorado-to-spend-15-million-yearly-to-protect/.

United Nations Development Program (UNDP). 2006. United Nations Human Development Report (UNHDR). Beyond Scarcity: Power, Poverty, and the Global Water Crisis.

United States Department of the Interior. Bureau of Reclamation (BOR), www.usbr.gov. Accessed 27 August 2009.

United States Department of the Interior, Bureau of Land Management (BLM), www.blm.gov. Accessed 27 August 2009.

United States Department of the Interior. U.S. Fish and Wildlife Service, Endangered and Threatened Species List, available at: http://ecos.fws.gov/tess_public/StateListing.do?state=all. Accessed 27 August 2009.

Utah Division of Water Resources. 2001. Utah's Water Resources: Planning for the Future, available at: http://www.water.utah.gov/planning/SWP/SWP2001/SWP_pff.pdf.

Water Strategist, 1991–2006. Available at www.waterstrategist.com

Chapter 9

Water options contracts to facilitate intersectoral trade

Joel Byrnes[1], Lin Crase[2] & Brian Dollery[3]
[1]KPMG, Australia
[2]La Trobe University, Melbourne, Australia
[3]University of New England, Armidale, New South Wales, Australia

Abstract

While much faith has been placed in the ability of market-based solutions to allocate water entitlements efficiently, relatively little effort has been made in fostering trade between urban and rural sectors. One barrier has been concerns regarding the future sustainability of rural communities following the trade of water out of rural areas. In this chapter, we demonstrate how the use of options contracts to facilitate temporary trade between the sectors may benefit both urban water utilities and water entitlement holders in irrigation districts.

Keywords: Intersectoral trade; Water options

1 Introduction

Creating the conditions for deep and liquid markets for the trade in water entitlements has been an enduring motif in the development of Australian water resource policy for at least the past 10 years (Crase *et al.*, 2007). Yet in terms of implementation, the focus has almost solely been on trade in rural water allocations. Trade from rural to other sectors has received relatively little attention due to at least two factors. First, in some locations considerable geographic barriers necessitate elaborate engineering solutions and considerable investment. Second, politicians and some rural communities have raised objections on the grounds that access to water is the lifeblood of rural Australia (Crase *et al.*, 2007) and therefore pivotal to continued sustainability of many rural communities.

In this chapter, we develop water options contracts previously proposed by Leroux and Crase (2007) and Page and Hafi (2007) for the temporary trade of water allocations between rural and urban sectors without requiring the transfer of ownership over the entitlement. We show that when purchased in relatively benign markets conditions, the instrument is an economically rational alternative to purchasing permanent entitlements.

This chapter itself consists of five additional parts. The background to the study area is provided in Section 2 to give the context for this chapter. In Section 3, a description of the general nature of options contracts is given, and we outline the applicability of this instrument to water markets. Existing papers that have investigated the potential for water markets in Australia are also reviewed in Section 3. The methodology followed, the data analysed and the results obtained are presented in Section 4. A discussion of some policy implications is given in Section 5, where we introduce the concept of water options contracts as a policy tool to return water to the environment. This chapter itself ends with brief concluding remarks in Section 6.

2 Potential for intersectoral trade in the Murrumbidgee Valley

Residential consumers in the Australian Capital Territory (ACT) have been subject to water use restrictions since 2002 (Pagan & Crase, 2008), and current government policy is to make this regime a permanent feature of life for the foreseeable future (Byrnes *et al.*, 2006). The fact that the Canberra water storage network suffers from a distinct deficit of capacity (Pagan & Crase, 2008) explains why a sizable portion of its water allocation is sent down the Murrumbidgee River as environmental flow (Edwards, 2007), while restrictions are simultaneously imposed on urban water use. It should not be a surprise that the ACT government has been one of the most aggressive in investigating the gamut of solutions to solve the urban water supply "crisis" that has prevailed in most of Australia's capital cities since the first few years of this decade. One of the "solutions" identified was the purchase of permanent water entitlements in the Tantangara Dam, located some 100 km upstream from Canberra on the Murrumbidgee River. Other options included the upgrade of existing dams, a more punitive set of water restrictions and the recycling of wastewater for potable use. In comparison, the "Tantangara option" has a number of advantages. First, Tantangara represents a sizeable storage capable of meeting the water demands of Canberra in most years. Second, the capital cost of the infrastructure to allow the transfer of water from Tantangara to one of Canberra's existing water storages is relatively cheap (estimated to be $38 million). Finally, the transfer of water from Tantangara would entail beneficial environmental flows (ACTEW, 2007).

Notwithstanding the advantages outlined above, the Tantangara option was rejected when first formally considered in 2005. This was essentially due to concerns surrounding less than certain property rights over entitlements to be held by the ACT urban water utility, stemming from the fact that Tantangara is located in the Snowy Mountains region of New South Wales (NSW), and subject to the control of the NSW government water bureaucracy (ACTEW, 2007). In essence, the ACT government feared that the transfer of water from Tantangara to meet the needs of urban water consumers in Canberra would be vetoed in times of extremely low storage levels, precisely when the water would be needed (Pagan & Crase, 2008).

However, a review of water supply options for the ACT in 2007 saw the Tantangara option revived. The plan would see high-security water entitlements purchased from those willing to sell (most likely irrigators from the Murrumbidgee and/or Coleambally irrigation districts) on a permanent basis. When required the water attached to the entitlement would be transported to a storage dam in Canberra, via either the Murrumbidgee River (a journey of approximately 100 km) or a newly constructed 20 km pipeline connecting the Tantangara to the Corin Reservoir. The plan calls for the purchase of rights to approximately 20 GL of water at an estimated cost of $30 million (ACT Government, 2007; ACTEW, 2007).

Implementation appears likely to fall flat for at least two reasons. First, there are non-trivial differences between the stipulation of water entitlement rights in NSW and the ACT (Pagan & Crase, 2008). Second, political objections to the permanent transfer of entitlements from rural to urban areas have been rapidly raised in other contexts, and seem just as likely to float to the surface in this instance (see Crase *et al.*, 2007 for a detailed discussion of this barrier to trade). Water options contracts potentially alleviate the second of these concerns.

3 The nature of options contracts and their applicability to water markets

An options contract gives the holder the right, but not the obligation, to undertake a defined action at a pre-determined future time, and at an agreed rate of exchange. When the holder of the contract opts to invoke the right, the option contract is said to have been exercised. There are two parties to an options contract: the writer and the holder. The writer of the contract is obligated to undertake an action (such as delivering an asset) in the event that the holder invokes her rights under the contract. Options contracts can be further classified into puts and calls. A call option gives the holder the right (but not the obligation) to purchase an asset, while the holder of a put option has secured the right (but not the obligation) to sell an asset (Jones *et al.*, 2007).

The value of options contracts derives from two distinct elements: intrinsic and time value. Intrinsic value is determined by the relationship between the physical asset and the price at which the option contract is exercised. An example helps to illustrate. Suppose a call option is entered into in order to purchase 1 ML of water for $300 ML^{-1} 3 months from today. Now suppose that on the "spot" or "physical" market, 1 ML of water trades for approximately $350 ML^{-1}. The intrinsic value of the option, therefore, is $50, since, if the option was to be exercised today, 1 ML of water would be purchased from the writer of the options contract for $300 and could be immediately sold in the spot market for $350, yielding a profit of $50 from the exercise of the option. Of course, should the contract specify that the option can only be invoked 3 months from today, a degree of uncertainty surrounds the intrinsic value, since the price in the physical market may fall below \$300 ML^{-1}. The degree of uncertainty essentially determines the time value of the contract. A relatively high degree of volatility in the price increases the time value of the option, while the opposite is true for lower degrees of volatility.

Options contracts have become an efficient means of transferring risk in financial markets, and although the basic premise of an options contract holds in their application to water markets, they differ in one key respect. Ownership is not transferred on exercise. Rather, the contract gives the right to temporary use of the water allocation. This brings two additional advantages. First, ownership of the entitlement remains with the writer, which may dilute some of the objections to intersectoral transfers. Second, the option can be exercised multiple times over a contract period, delivering some supply security for the urban water utility.

Given the special nature of water options contracts, Michelson and Young (1993), who were the first to propose options as a means of transferring water between urban and rural sectors, identified a number of other features that must be incorporated. First, water supply in the region in which a transfer was to occur must be of sufficient reliability to provide water to the urban authority in drought years, since if this were not the case the option would not meet its ultimate objective: the provision of fungible water to be distributed by the authority. Second, the legal definition of property rights to both water entitlements and allocations should be definable and capable of transfer between the parties to the contract.

Third, since it is unlikely that an irrigator would agree to enter into an option that placed an obligation on him to give up "essential" water in a drought year, the agricultural enterprise of the seller of water must be capable of being suspended for a period of time. It follows that while a typical viticulturalist would not be in such a position, a cotton irrigator would be potentially well placed to enter into the contract. Fourth, each party to the contract must have sufficient knowledge of the present and future values of water allocations and entitlements in order to form reasonable expectations. Clearly, markets where opacity with regard to volumes and prices exists would not be well suited to the deployment of water options contracts.

Fifth, since the underlying value of an options contract to both the buyer and the seller is a function of the probability of the option being exercised, sufficient information must be available to allow reasonable estimates of the probability of exercise. Finally, in order for water options to be a realistic alternative for urban water authorities for securing urban water supplies during periods of drought, the total costs to be incurred in entering into and exercising the option must be less than the urban water authorities' next most costly supply alternative (examples include outright purchases of water entitlements and various engineering solutions such as dam augmentation and the construction of a desalination plant).

In terms of the operational aspects of water options contracts, Michelsen and Young (1993) argued that the exercise price under the option should be anchored to the economic value added per megalitre of water applied in the given agricultural pursuit (e.g. gross margin per megalitre). To standardise water options contracts to the extent that permitted trade, the quantity of water to be transferred, the time of delivery and method of delivery were each required to be specified in the terms of the contract. Michelsen and Young (1993) also recognised that since the proposed contract had a life of between 5 and 10 years, ex-ante arrangements should be in place to allow the owner of the underlying entitlement to sell the entitlement, without unduly disadvantaging the purchaser of the option. This could be achieved by providing first right of refusal to the purchaser of the option contract in the event that the entitlement was listed for sale.

3.1 Valuing water options contracts

There are two steps to determine whether a water options contract has value to the holder. In the first, the capital cost of obtaining the next least cost alternative supply source is compared to the cost of exercising the option. In the case that exercising the contract is a cheaper alternative, the option has economic value. In the second step, the cost of purchasing the contract is compared to the value of holding the contract. Should the premium payable not extinguish the value of holding the contract, the urban water utility would benefit from purchasing the option.

3.2 Previous efforts at exploring water options in Australia

As noted previously, it is generally acknowledged that the literature on water options contracts began with Michelsen and Young (1993), who proposed the use of a highly stylised variant of plain vanilla options contracts to facilitate the movement of water between rural and urban water markets. Michelsen and Young (1993) established that in the case of Southern California, the purchase of a water options contract by an urban water authority from irrigators upstream was a least-cost method to ensuring access to water in the case of drought. A detailed outline of the model employed is presented in Section 4.1 below.

Ranjan *et al.* (2004) addressed the likelihood of various potential parties to water options contracts perceiving sufficient reward from the risk incurred through entering into the contract. The authors concluded that while the existence of a derivative market in parallel with the "physical" market was useful in terms of reducing uncertainty, the risk to farmers of being obligated to supply water at a time of drought was substantial, limiting the growth in options markets for water products.

The embryonic empirical Australian literature on water options began with Page and Hafi (2007) who investigated the potential for water options contracts as a means of providing a financial incentive for irrigators to allow "excessive" water allocations to flow downstream to simulate natural flood events. Although in the context of the recent drought in the south-east of Australia the concept seems remote, such an options contract provides an avenue for environmental managers to reveal a value for environmental flows, and for irrigators to reveal the value at which they would consider selling water for the environment. To measure the economic benefits to the environmental manager from holding a water options contract, Page and Hafi (2007) employed the method proposed by Michelsen and Young (1993). Under the majority of scenarios tested, the option was shown to provide a positive economic value.

Page and Hafi (2007) also made use of the Michelsen and Young (1993) method to investigate the economic value to an urban water authority from holding a water options contract as an alternative to entering into more traditional approaches to augmenting urban water supply. Once again the authors established that under the majority of scenarios tested, the urban water authority would derive a positive net present value from entering into an options contract when compared to the alternatives.

More recently, Leroux and Crase (2007) applied the Michelson and Young (1993) model to an examination of the potential for water options contracts to be agreed between irrigators upstream of the Australian regional town of Wangaratta in Victoria as an alternative to the outright purchase of water entitlements. In summary, they found that for a range of scenarios it may be worthwhile entering into water options contracts since there was a net present value from doing so. However, the authors noted that this was conditional on the premium demanded by the writer of the contract.

Rather than investigating the potential for water options contracts, Watters (1995) made use of the Black–Scholes (BS) derivative pricing method to value existing call options in the southern California water market. Watters (1995) established that the options contracts were generally priced at so-called "fair-value", although reflecting a high probability of exercise, suggesting little risk transference, was actually taking place.

The underlying price profile of water allocations in most rural water markets does not follow a classic normal distribution. As a result, the use of the BS pricing method has been drawn into question by a number of authors, since the BS method has as an assumption normality in the distribution of the price of the underlying asset. Villinski (2003) sought to overcome this limitation by pricing the option financially via dynamic programming.

Williamson *et al.* (2007) investigated the potential for water options contracts to be agreed between irrigators, with the aim being to facilitate the transfer of water from relatively low-value irrigation enterprises to relatively higher-value applications. The authors made use of the Michelson and Young (1993) model, which made a substantial contribution to the literature by estimating the premium payable to the writer of the contract under a number of scenarios. The valuation was undertaken using an amended version of the BS model that took into account the non-uniform distribution of water allocation prices. Williamson *et al.* (2007) found that there were few scenarios in which an options contract between irrigators would be viable, essentially as both parties had largely similar demand preferences during periods of drought.

4 Methodology, data and results

In this section, we contribute to the literature by both estimating the present value of benefit (PVOB) to the ACT water authority from holding the option and the premium that would be required by the writer of the contract, via the BS pricing model.

4.1 Estimation of the PVOB from holding a water option

Following Michelson and Young (1993), Leroux and Crase (2007), Page and Hafi (2007) and Williamson *et al.* (2007), the PVOB from holding the option contract is estimated as follows.

$$\text{PVOB} = \Sigma^{T}_{t=0}\,[(\text{K}_{t=0}{*}r + M_t - B(1{-}P)){-}(E{*}P)_t]\text{d}_t + [\text{K}_{t=0}{-}\text{K}_{t=0}(1 + a)^T]\text{d}_t \qquad (1)$$

where
PVOB = net present value of option benefits
t = year
T = contract termination year
$\text{K}_{t=0}$ = capital cost of alternative supply at the beginning of the option term
r = annual interest rate
M= annual maintenance cost of the alternative
B = price per megalitre of temporary water sales
E = exercise cost of option
P = annual probability of option exercise ($0 \leq P \leq 1$)
d_t = discount factor for present value, $1/(1 + r)^t$
a = annual rate of appreciation of alternative supply.

In essence, eqn 1 compares the capital cost of obtaining the alternative supply and any annual costs and/or benefits associated with obtaining the alternative supply ($\text{K}_{t=0}{*}r + M_t - B(1{-}P)$), to the cost of exercising the option contract ($E{*}P$). The present value of future transactions is obtained by applying a standard discounting factor (d_t). A PVOB is estimated for each of the contracts bundled into the multiple exercise contract. Thus, in the case of a 10-year contract, 10 PVOBs are estimated and summed to provide the PVOB of holding the contract over 10 years. The second term, $[\text{K}_{t=0} - \text{K}_{t=0}(1 + a)^T]\text{d}_t$, allows for inclusion of appreciation or depreciation in the value of the alternative supply, discounted to allow valuation at present value. It is important to note that the second term is only considered in the final year of the contract.

4.1.1 Data

In this example, we estimate the value of contracts spanning both 5- and 10-year periods. The alternative supply to be secured is the purchase of 20 GL of high-security water entitlements on the permanent market. As outlined in Section 2, investigations by the ACTEW suggested the cost of this purchase would be approximately \$30 million, which equates to \$1,500 ML^{-1} (ACTEW, 2007).

Since data regarding trades on the permanent market in the relevant district are scarce, the estimate by ACTEW was taken on face value.

The annual risk-free interest rate was arbitrarily set at 5% and left constant; however, it is acknowledged that this need not be the case. The annual maintenance cost in this instance consisted of payments that would be required by the Snowy River Scheme for lost hydroelectricity generation, estimated at \$270 ML^{-1}. We assume that unused portions of any alternative supply could be sold into the temporary water market at \$300 ML^{-1}, and that the probability of making temporary sales is given by $(1–P)$.

Following Michelson and Young (1993), Leroux and Crase (2007) and Williamson *et al.* (2007), the exercise price was taken as the gross margin per megalitre of water consumed in the irrigation of three separate crops in the Murrumbidgee Valley: barley, faba beans and lucerne. Data were sourced from NSW Department of Primary Industries gross margin budgets for 2007–08 (Pagan & Crase, 2008), and act as the "base" scenario 1. Two further scenarios were modelled in which gross margins were deliberately increased.

The annual probability of exercise is based on modelling by ACTEW (2007) that suggests water restrictions are likely to be imposed every 3 years out of 10. We leave the rate of appreciation at zero for the sake of simplicity.

4.1.2 PVOB from holding water option contract

Table 2 presents the estimates of the PVOB to the urban water utility from holding the water options contract written by the three different irrigators, under each of the three scenarios and for contract lengths of 5 and 10 years.

A number of patterns are apparent. First, the value of holding a 10-year contract is universally higher than that of holding the 5-year contract, regardless of who the contract is purchased from. Second, the higher the exercise price of the contract, the lower the present value of holding the contract. This is intuitively appealing, since the economic principle underpinning water options is that water will move from a relatively lower-value use to a higher value. Combined, the results suggest an economic benefit (in present terms) to the Canberra urban water utility of entering into either of the two contracts. This is consistent with the findings of Page and Hafi (2007).

Table 1: Gross margin assumptions.

Crop	Gross margin: first scenario	Gross margin: second scenario	Gross margin: third scenario
Barley	48.75	100	150
Faba beans	127	300	350
Lucerne	163	370	450

Table 2: Netpresent value of benefits from holding option contract under three scenarios.

PVOB	Barley		Faba beans		Lucerne	
	5 years	10 years	5 years	10 years	5 years	10 years
Scenario 1	\$521	\$929	\$420	\$748	\$373	\$665
Scenario 2	\$455	\$811	\$195	\$347	\$104	\$185
Scenario 3	\$390	\$695	\$130	\$232	\$32	\$58

4.2 Estimation of premium payable to writer of option contract

The BS approach to determining the "fair value" premium payable to the writer of an options contract is now almost universally employed in the pricing of financial options (Jones *et al.*, 2007). Although the underlying nature of water prices violates a basic assumption (normality), we make use of this approach because of its general acceptance. The results are obviously limited by this modelling choice.

In essence, the BS model prices an options contract on the likelihood that the contract will be exercised, with higher probabilities requiring relatively higher premiums. It follows that the relative values of spot and exercise prices are of crucial importance. The BS approach is specified as follows:

$$c = S_0 N(\mathrm{d}_1) - K\mathrm{e}^{-rT} N(\mathrm{d}_2) \tag{2}$$

and

$$\mathrm{d}_1 = (\ln(S_0/K) + (r+\sigma^2/2)T)/\sigma\sqrt{T} \tag{3}$$

$$\mathrm{d}_2 = \mathrm{d}_1 - \sigma\sqrt{T} \tag{4}$$

where c is the European call price, S_0 is the stock value at time zero, N is the cumulative normal distribution, K is the strike/exercise price, r is the risk-free rate of interest, σ measures price volatility and T is the time to maturity of the option in years.

4.2.1 Data

In the case of water option contracts, the stock value (S_0) is the price of temporary trades in high security water, expressed in dollars per megalitre. We make use of a data set supplied by the Murrumbidgee Horticultural Council (2007) over the period 2006–07. A feature of the data set is the relative lack of depth in trades and the extreme volatility from mid-2006 to late 2007. The series is presented in Figure 1, with both volume and spot price displayed.

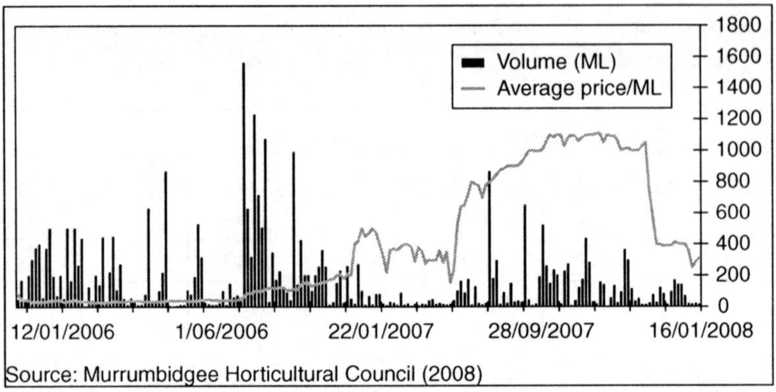

Figure 1: Average price and volume of trade in Murrumbidgee Valley. Adapted from Murrumbidgee Horticultural Council (2008).

K and r are as previously defined. Volatility in spot prices is clearly greater than 1, evident from the range of prices (\$29–1401 ML^{-1}). Under the bold assumption that a degree of stability will return to the market as the current drought breaks, we have imposed a volatility term of 0.5. As outlined above, contracts of two maturities (5 and 10 years) are specified.

If the option was in the money, the urban water utility would compare the value of holding the option with the cost of purchase, represented by the premium. When the contract is out of the money, it is left unexercised, even though the urban water utility foregoes the premium, since water can be sourced from the spot market at a price less than the exercise price. In Table 3, the range of spot water prices at which there is still a surplus PVOB after having deducted the

Table 3: Spot prices at which PVOB is greater than premium payable.

Scenario	Barley	Faba beans	Lucerne
5-year contract			
1	\$70.31–525	\$130–450	\$175–416
2	\$100–525	Premium > PVOB	Premium > PVOB
3	\$151.5–475	Premium > PVOB	Premium > PVOB
10-year contract			
1	\$70.31–955	\$130–777.50	\$175–416
2	\$100–843	\$300–402.00	Premium > PVOB
3	\$151.5–750	Premium > PVOB	Premium > PVOB

premium is reported. The lower value is by definition the spot price at which the contract is in the money. This is reported for each of the three crops and each of the three gross margin scenarios.

4.3 Results

A generally applicable principle is evident in Table 3. Those producing crops that add the least economic value per megalitre of water are most likely to trade with the urban water authority via an options contract. This stems from the fact these farmers require a relatively lower exercise price, resulting in relatively higher PVOB surpluses after the payment of premiums.

5 Policy implications

The results presented in this chapter suggest it would be sensible for the urban water utility to enter into an options contract when it was not facing an immediate supply constraint due to drought, but when prices were relatively low, since the higher the spot price of water relative to the exercise price, the higher the compensation required by the writer of the contract. Thus, in periods of relatively high water prices, options contracts are less likely to be entered into by the urban water utility. The premium paid to the writer in times of supply security represents the price of obtaining future supply security well before a "crisis" is encountered. Some may regard this as money well spent. The rash of expensive infrastructure projects being fast-tracked by state governments around Australia demonstrates the cost of taking a "just in time" approach to urban water planning. This would also be beneficial to the farmer since he/she receives income from selling the option at a time when the option is unlikely to be exercised.

There is also scope for the concept of a water options contract to be employed by regulatory bodies in their efforts to reallocate fungible water for environmental flows. This is particularly so in the current Australian context, where the Commonwealth government has turned to "buying back" water entitlements from irrigators to increase environmental flows; a policy stance that is meeting considerable opposition in rural communities. Through entering into water options contracts with irrigators, the Commonwealth may overcome many of those objections, since entitlements are not being permanently removed from the community.

In an operational sense, the government could purchase an option to buy *allocated* water from an irrigator. The trigger for the option becoming active might be a given stream inflow reaching a specified minimum level (30% lower than normal, for example). The government would then give the irrigator at least 9 months notice that they intend exercising the option, allowing the irrigator to make the necessary adjustments.

Should the option be triggered, the water allocated against the irrigator's entitlement would be delivered to the river and left to generate a minimum flow in the river. The irrigator would be paid a price per megalitre, with that price agreed between the parties in advance, giving both parties price certainty. The irrigator would only be bound to deliver the allocation against his/her entitlement that had been announced by the water planner, removing the risk that the farmer would need to enter the spot market to deliver a specified volume.

Regardless of what happens to rainfall over the life of the contract, the irrigator would be paid a premium by the government. In wet years, when the government has not exercised the option, the irrigator would farm the land as usual, while in dry years the irrigator would be paid income not to make use of the allocation. Perhaps of most appeal to both the government and irrigators, the government could potentially avoid paying large sums of money for water entitlements that in dry years may not contribute much by way of fungible water to the river. Most importantly, however, ownership of the water entitlement stays with the irrigator, potentially contributing to the long-term sustainability of rural communities.

6 Conclusions

In this context at least, there would appear to be value for the urban water utility in the purchase of water options contracts. However, the results suggest that the transaction is likely to be with those irrigators adding relatively little value per megalitre of water. This is intuitively appealing, since it sits well with the basic premise of water trading: trade will move water from relatively low- to higher-value uses. Yet as is so often the case in water policy, options contracts do not represent a universal or perfect solution to the thorny issue of intersectoral trade. Furthermore, while the results presented here are promising, a number of important caveats must be considered. Property rights must be certain. Planning must be such that there is a high degree of certainty surrounding the proportion of the applicable resource available. Knowledge of interactions between basins and the consequences of moving allocations from one part of the system to another must be well advanced so that the parties can be reasonably certain of the quantity of water required to make the delivery.

References

ACT Government. 2007. Water supply security fact sheets. http://www.thinkwater.act.gov.au/more_information/publications.shtml#water_security. Accessed 27 August 2009.

ACTEW Corporation Ltd. 2007. *Water Security for the ACT and Region: Recommendations to the ACT Government*. ACTEW, Canberra, Australia, 2005.

Byrnes, J., Crase, L. & Dollery, B. 2006. Regulation versus pricing in urban water policy: the case of the Australian National Water Initiative. *The Australian Journal of Agricultural and Resource Economics,* 50: 437–449.

Crase, L., Byrnes, J. & Dollery, B. 2007. The political economy of urban–rural water trade. *Public Policy,* 2(2): 130–140.

Edwards, G. 2007. Urban water management. In: Crase, L. (Ed.), *Water Policy in Australia: The Impact of Change and Uncertainty*. RFF Press, Washington, DC, pp. 144–165.

Jones, C., Shamsuddin, A. & Naumann, K. 2007. *Investments Analysis and Management*, 2nd edition. John Wiley & Sons, China.

Leroux, A. & Crase, L. 2007. Options Contracts for Managing Inter-Sectoral Water Trade: A Preliminary Investigation of the Feasibility of Urban-Irrigation Options Contracts in the Ovens Basin, Victoria, Australia. *Proceeding from Australian Agricultural and Resource Economics Society Annual Conference*, February, Queenstown, New Zealand.

Michelsen, A. & Young, R. 1993. Optioning agricultural water rights for urban water supplies during drought. *American Journal of Agricultural Economics*, 75: 1010–1020.

Murrumbidgee Horticultural Council. 2007. High Security Water Entitlement. http://www.murrumbidgeewater.com.au/index.php?id=6&no_cache=1. Accessed 27 August 2009.

NSW Department of Primary Industries. 2007. Farm enterprise budget series: Gross margin budgets. www.agric.nsw.gov.au. Accessed 27 August 2009.

Pagan, P. & Crase, L. 2008. Institutional Constraints and Inter-Jurisdictional Water Trade: The Case of the Australian Capital Territory and New South Wales. *Proceedings of the 51st Annual Conference of the Australian Agricultural and Resource Economics Society*, February, Canberra, Australia.

Page, S. & Hafi, A. 2007. *Urban Water Options Contracts: Rural to Urban Water Trade*. Australian Bureau of Agricultural and Resource Economics, Canberra, Australia.

Ranjan, R., Gollehon, N. & Aillery, M. 2004. *Explaining Participation in Spot and Options Markets for Water*. International Agriculture Trade Policy Centre, University of Florida, Gainesville, FL, USA.

Villinski, M. 2003. *A Methodology for Valuing Multiple-Exercise Options Contracts for Water*. Centre for International Food and Agricultural Policy, University of Minnesota, Minneapolis, MN, USA.

Watters, P.A. 1995. *Efficient Pricing of Water Transfer Options: Non-Structural Solutions for Reliable Water Supplies*. University of California, Riverside, California, USA.

Williamson, B., Villano, R. & Fleming, E. 2007. Structuring Exotic Options Contracts on Water to Improve the Efficiency of Resource Allocation in the Water Spot Market. *Proceedings of the 51st Annual Conference of the Australian Agricultural and Resource Economics Society*, February, Canberra, Australia.

Chapter 10

Assessing alternative models for farmers' ability to pay for irrigation water

Robert A. Young[1], Susanne M. Scheierling[2] & Richard L. Bowen[3]
[1]*Emeritus, Department of Agricultural and Resource Economics, Colorado State University, USA*
[2]*Energy, Transport and Water Department, The World Bank, USA*
[3]*Department of Natural Resources and Environmental Management, University of Hawaii at Manoa, USA*

Abstract

To sustain public projects for irrigating agricultural crops over long periods requires expenditures for investment, operations and maintenance, and periodic rehabilitation. Governments have historically been the primary sources for financing these projects, but farmer beneficiaries are increasingly expected to contribute substantially toward recovering costs. Estimates of farmers' ability to pay (ATP), or repayment capacity, for part or all of the costs of irrigation water supply facilities are useful in deciding on how much of these costs farmers can and should pay. Three alternative models for determining farmers' ATP have been identified. In this chapter, we formalize these alternative models to defining "ability to pay," evaluate them, and illustrate their strengths and limitations – both from theoretical perspectives and with an empirical analysis based on data from a case study in the Kyrgyz Republic. We find that in the study region, the alternative models yield significantly different estimates of ATP. The model that we regard as theoretically correct yields a significantly lower ATP than does the one embodying the implicit approach of study-area non-specialists and non-economists.

Keywords: Ability to pay; Charging; Cost recovery; Irrigation; Repayment capacity

1 Introduction

Throughout the world, irrigation of agricultural crops with surface water is mainly carried out in large public schemes or projects. They typically consist of networks of facilities for capture, storage, and conveyance of water from the point of capture to numerous individual farmers' plots.

For their productivity to be sustainable over the longer term, these projects require substantial expenditures on initial development, recurrent operation and maintenance activities, and occasional major rehabilitation to avoid deterioration of structures and keep the technology updated. While financial arrangements to fund irrigation projects usually contemplate government and/or donor sources, farmer beneficiaries are increasingly expected to provide a substantial share of the costs. Cost recovery from beneficiaries will help provide funds for adequate current financing of the irrigation system, for retirement of loans that funded the investment, or for future public needs. Accordingly, assuring the sustainability of irrigation schemes typically requires some formal analysis on which to base the share of irrigation system financing that farmers can support (their "repayment capacity," or so-called "ability to pay" (ATP)), versus the financing provided by government or donor agencies.

This chapter reports our conceptual and empirical evaluation of several alternative models that have been used to analyze and evaluate farmers' ATP. Our analysis was in part motivated by experience with an Asian Development Bank-funded study of ATP in the Kyrgyz Republic[1] (Asian Development Bank, 2007). A diversity of opinions on how to estimate ATP was observed among those involved in planning and carrying out that study. Below, we first develop a conceptual framework to distinguish alternative approaches, then apply this framework to a case study of a region in the Kyrgyz Republic and, finally, draw conclusions for future directions in assessing farmers' repayment capacity.

2 Background and conceptual framework

In this section, we review the policy background and general economic framework for analyzing the recovery of costs of irrigation projects, introduce some terminology, and present a formal modeling structure for determining irrigators' ATP.

2.1 Policy background

To begin, it is useful to recall O'Mara's (1990) two polar perspectives on how public irrigation projects may be financed. One type he calls the *fiscal autonomy* model, which reflects the typical approach to other network services such as electricity or telecommunications. The fiscal autonomy model assumes the existence of a well-defined, specialized, and autonomous organization that supplies specified services to its clientele and is financed all or mainly by the payments of its clientele for services received. The second, polar approach is called the *fiscal dependence* model. It views governments, because of the special economic nature of agricultural production, as naturally choosing to intervene in the agricultural sector. Given the uncontrollable and unpredictable fluctuations of climate, weather and economy, and low short-run elasticities of both supply and demand for farm products, individual farm households tend to experience widely fluctuating output prices and incomes for reasons beyond their control.

Moreover, many water users are financially unable to pay the full cost. In addition to agricultural income maintenance and stability, food security and employment may be other policy concerns. In this view, irrigation is thought of as the only one among many governmental activities directed toward supporting and stabilizing the agricultural sector, and a considerable portion of irrigation financing is expected to come from public sources.

2.2 General economic framework

For discussions of repayment capacity, it is useful to distinguish among several important economic concepts: price, charge, value, and cost. *Price* generally refers to a monetary payment per unit volume (e.g., cubic meter) of water. In a market context, price reflects the willingness to pay (WTP) of the marginal buyer and the willingness to sell of the marginal supplier, so price and value are synonymous. In a nonmarket context, almost always the relevant situation in water policy evaluations, *value* usually means a synthetic estimate of beneficiaries' WTP (also called a shadow price). *Costs* may refer to the outlays necessary to purchase some goods or services. More generally, they are the value of the opportunities foregone because of the commitment of resources to an activity, or sometimes the WTP to avoid detrimental effects. The economic theory of pricing in public sector contexts is mostly cost-based.

A further distinction is between determining what to charge for water and the economic valuation of water. In the context of water policy, *charging* (or tariff-setting) refers to the process of setting the monetary amount of costs to recover from water users for purposes of sustaining the supply agency and its tasks, with possible consideration given to providing incentives for water conservation and assuring supplies for the less well-off. *Economic valuation*, in contrast, is the application of nonmarket economic methods to determine the economic value, based on the concept of WTP for water services, for purposes of evaluating proposed management and allocation policies. Charging and economic valuation reflect the opposite sides of supply and demand forces, respectively, that lead to equilibrium for market goods and services. In nonmarket contexts, charging represents the important policy choice of how much of costs to recover, and how and from whom to recover them. Charging mostly reflects supply-side considerations, primarily those based on costs. Demand-side considerations, those related to WTP at varying quantities, are reflected in economic valuation processes. Economic valuation is a study of water users' behavior and of what priorities people are expected to place on water management policies. In the context of water policy choices, nonspecialists sometimes incorrectly equate the observed price or charge with economic value. Valid selection of policy instruments regarding water must reflect both cost and value considerations, with the level of charges limited by the water users' ATP, rather than being defined by it.

A rather extensive literature on financing, pricing, and cost recovery for irrigation water projects has emerged in recent years. These writings include Cornish *et al.* (2004), Dinar (2000), Molle and Berkhoff (2007), Sampath (1992),

Small and Carruthers (1991), and Tsur *et al.* (2004). Even so, relatively little attention has been directed at formally conceptualizing and measuring farmers' ATP or repayment capacity, although exceptions (which will be discussed later) are found in Gittinger (1982) and Moore (1999).

A major consideration in setting up a plan for irrigation cost recovery is the organization and configuration of the irrigation water delivery system. The typical design, particularly in Asia, is for large numbers of smallholder farmers to be supplied by an extensive network of large publicly operated canals and smaller distributaries. The water distribution system is typically comprised of unlined earthen canals and ditches, which leak extensively and lack accurate measurement devices. Of the several potential methods or forms of charging for irrigation water, the most common one is *area-based* (Tsur *et al.*, 2004). In this form, a flat annual charge is set according to the land area (e.g., in hectares) being irrigated. Area-based charges, however, are criticized for the lack of any incentive to conserve water by reducing the number of irrigations, changing the amount of water applied per irrigation, or changing to less water-intensive crops. The main alternative to an area-based charge is some sort of *volumetric* charging system. Such an approach, of course, requires some agreed-upon method of measuring volume and the capacity to monitor deliveries to each smallholder plot. Our study method is most applicable to the most frequently encountered case: area-based charging, but could be adapted to volumetric charging approaches with a more complex mathematical programming model.

2.3 General definition of ATP

In economic terms, ATP can be defined as the maximum amount a representative farmer can be expected to be able to contribute toward financing the costs of some specific plan to develop, operate, maintain, and/or rehabilitate irrigation infrastructure under assumed or forecasted technology, policy, and market conditions. This maximum amount the farmer can pay for water supply is the amount of income remaining after all the other resources have been paid (in the amounts required to draw them into production, measured by their market price or, for nonpurchased inputs, by opportunity costs). The concept of ATP is important in irrigation planning and cost recovery analysis because it provides an approach to establishing appropriate farmers' shares of costs incurred in developing, operating and maintaining, and rehabilitating facilities for delivering irrigation water.

As will be shown below, this definition of ATP is similar – but not always identical – to the definition for WTP, the standard expression for economic benefits of producers' goods in cost–benefit analysis.

2.4 Conceptual model of ATP

2.4.1 Basic model

To enhance the clarity of our discussion, we develop a more formal model of ATP. It is derived from the model of net economic benefits from irrigation, or WTP. WTP is conventionally based on the theory of the firm, which describes the basic production-side decision-making unit in mainstream microeconomic theory. (The formulation adopted here is developed in more detail in Young (2005, Chapters 3 and 5) The firm is understood to be a legal entity (such as a proprietorship, partnership, or corporation) that owns some inputs (also called factors of production) and purchases others, transforming these inputs into outputs of goods or services. The standard model begins with a *production function* – a relationship that shows the maximum production that can be obtained from all possible combinations of given inputs at the given state of technical knowledge – that serves as the technical description of the firm. In symbols, the production function for irrigated crop production can be expressed as follows:

$$Y = f(X_M, X_H, X_K, X_L, X_W, C, E) \tag{1}$$

where X stands for the quantity of an input and Y refers to the quantity of an output. The subscripts M, H, and K refer to inputs that are typically purchased by the farm (in the theory texts, usually termed as "contractual" inputs) – where M is materials, energy, and equipment, H labor, and K (borrowed) capital. The capital and operating costs of the farm's water distribution system (ditches, pipes, sprinklers, and the like, and the energy to operate them) are typically treated as part of the materials, energy, and equipment costs. The remaining inputs are assumed here to be owned or "noncontractual," although some of them often may also be purchased. The owned inputs are specialized inputs, those whose prices, in reality, are determined after the fact by the outcomes of managerial decisions within the existing natural and economic environments, but in water valuation practice must be estimated *ex ante* by opportunity costs. The subscript L refers to land, C to equity capital of the firm, and W to irrigation water. (Note that water is treated as one of the owned inputs.) E stands for opportunity costs of owned skills, management, technical knowledge, and entrepreneurial creativity.

This model is written to be as fully general as possible, although it may be overly complex for small farms growing staple crops in developing countries. We deem it important that the production function be described here in as much detail as possible. Part of our subsequent critique of some models for estimating ATP is based on the belief that such models are mis-specified, in the sense that some inputs are omitted, and therefore farmers' costs are understated and ATP overstated.

For this general case, the firm is assumed to operate in properly functioning factor and product markets, and faces perfectly elastic supplies for factor inputs and perfectly elastic demand for its output, so that prices are known and constant.

Using what is called the "residual method," we next move from the production function to the rent function for water. The term "rent" in economic theory differs from the meaning ascribed in everyday language: it refers to nonobservable income imputed to an input in limited supply. The residual method is used primarily for valuing nonmarketed producers' goods (also called intermediate inputs). Given the "adding-up theorem" – which asserts that the sum of all the costs of inputs exactly equals the total revenue from production – the net producers' rent attributable to a nonpriced productive input is found by subtracting all other estimated costs of production from forecasted total value of output. That is, the remaining value is assigned to the nonpriced residual input, in this case, water.

In eqn (2), let *R* represent economic rents to the residual claimant, and the superscript *W* stand for water. *P* refers to the array of prices of outputs and factor inputs. By convention, the net rent formulas are usually standardized in terms of land, that is, expressed in per unit land (e.g., per hectare). Here, as is also typical, these formulas are expressed in annual (rather than present value) terms. Assuming input and output prices and production technology remain constant, and durable input costs are expressed in annual equivalent terms, the basic annual water-related rent formula for a single commodity can be written symbolically as follows:

$$R^W = (Y \times P_Y) - [(P_M \times X_M) + (P_H \times X_H) + (P_K \times X_K) + (P_L \times X_L) + C + E] \quad (2)$$

Typically, the non-water inputs amount to the major share in total value of output, and the calculated residual rent, while not trivial, is a relatively small portion of it. In consequence, the estimated residual can be very sensitive to changes in the non-water costs, and particularly sensitive to omissions of any cost components.

Note that this formulation represents a Marshallian long-run, partial equilibrium model. Our assumption of the long-run model, in which all factor inputs are assumed to be variable, is important to our results. Since it is assumed that each of the non-water inputs are paid their marginal value product, the rent formula represents the measure of a representative farm firm's long-run WTP for water for a crop on a unit land area (Young, 2005, Chapter 5). (A pervasive challenge in estimating both WTP and ATP for irrigation water is that the non-water-owned inputs *C* and *E* in eqn (2) are often themselves nonpriced, and must be estimated by their opportunity costs or by rules of thumb.)

2.4.2 Alternative ATP models

ATP calculations are normally conducted as part of or as a follow-up of the economic benefit estimation in cost–benefit analysis of some plan or proposal to invest public funds for developing, repairing, or renovating an irrigation distribution network. We now show how the calculations for net economic benefit, or WTP, provide the basis for estimating ATP. Depending on the situation, ATP may or may not equal WTP for irrigation water. Consider three possible cases of interest:

Model 1: This case is based on the following assumptions: Farmers operate within properly functioning competitive factor and product markets. Specifically, opportunity costs of input factors – for example, labor, materials, or capital – equal their market prices. Further, social values of products equal their market prices. There is no public intervention into either factor and/or product markets. These assumptions are embodied in eqn (2), rewritten as eqn (3):

$$\begin{aligned} \mathrm{WTP}^W(1) = (Y \times P_Y) - [(P_M \times X_M) + (P_H \times X_H) \\ + (P_K \times X_K) + (P_L \times X_L) + C + E] \end{aligned} \quad (3)$$

As defined earlier, ATP^W refers to the maximum amount the (representative) farmer can be expected to be able to contribute toward financing the costs of some specific plan to develop, operate and maintain, and/or rehabilitate irrigation infrastructure under assumed or forecasted technology, policy, and market conditions. The net ATP equals the rents or returns available to pay for water after all other inputs are paid the amounts necessary to draw them into production. Thus, the Model 1 formula for ATP^W is

$$\text{ATP}^W(1) = (Y \times P_Y) - [(P_M \times X_M) + (P_H \times X_H) + (P_K \times X_K) + (P_L \times X_L) + C + E] \quad (4)$$

The right-hand sides of eqns (3) and (4) are identical, both satisfying this definition. In other words, in an economic environment of properly functioning competitive factor and product markets, and where the economic analysts judge that no shadow pricing is needed in either factor or product markets, $\text{ATP}^W(1) = \text{WTP}^W(1)$.

Model 2: The changed assumption for this case is that prices of one or more relevant factors and/or products are not determined in properly functioning markets. There may be public intervention into factor and/or product markets, a situation frequently encountered in the agricultural sector of many countries. In the cost–benefit analysis of these cases, one or more of factor and/or product prices have been obtained by shadow pricing. In the prototypical example of Model 2, the social opportunity costs of labor are estimated to be less than the market wage faced by farmers. Cost–benefit studies in developing countries with unemployed workers typically calculate social WTP with a shadow wage assumed to be less than the market wage. In the extreme case, the opportunity cost (shadow wage) of labor can be assumed to be zero. In eqn (5), letting $(P_H \times X_H)$ fall to zero, we have $\text{WTP}^W(2)$ as follows:

$$\text{WTP}^W(2) = (Y \times P_Y) - [(P_M \times X_M) + (P_K \times X_K) + (P_L \times X_L) + C + E] \quad (5)$$

Other things equal, for this example, the residual $\text{WTP}^W(2) > \text{WTP}^W(1)$ by the amount that labor cost are reduced via shadow pricing. With farmers still paying the market wage for labor, the specification of ATP^W is the same as in eqn (4):

$$\text{ATP}^W(2) = (Y \times P_Y) - [(P_M \times X_M) + (P_H \times X_H) + (P_K \times X_K) + (P_L \times X_L) + C + E] \quad (6)$$

Thus, reflecting differing assumptions about the market price versus social opportunity costs of some inputs, $\text{ATP}^W(2)$ differs from $\text{WTP}^W(2)$. In the above example, where the social opportunity cost of labor is estimated to be less than the market wage, $\text{ATP}^W(2) < \text{WTP}^W(2)$.

Other examples of Model 2 arise when crop prices received by farmers do not reflect their social value. For example, in the United States, price support programs lead to prices faced by farmers being higher than their social value – as with cotton and wheat. Conversely, some developing countries have held crop prices down to favor urban interests, leading to crop prices that are less than estimated social values. The first of these instances leads to an $ATP^W(2) > WTP^W(2)$. The second instance leads to $ATP^W(2) < WTP^W(2)$.

Model 3: This case assumes a different formula for defining ATP^W. Within the economic/policy conditions of either case 1 or 2 above, for case 3, maximum $ATP^W(3)$ is assumed to equal "net farm income," that is, returns to water *and* all other owned inputs. We have not come across any formal definition of ATP^W of precisely this form in the literature. However, this case reflects the informal, intuitive concept non-economists and even nonspecialist economists proposed in our discussions of appropriate modeling of ATP^W in the Kyrgyz Republic. This formulation is similar to models of private short-run farm decision-making (such as for annual crop land allocation) emphasized in textbooks on farm management (e.g., Kay *et al.*, 2007), and thus is familiar to analysts trained in modeling private farm decision-making, but not necessarily familiar with social cost–benefit analysis. In this case, ATP^W can be written as in eqn (7), where the residual is defined as revenue minus the costs of purchased inputs, M, labor, H, and borrowed capital, K:

$$ATP^W(3) = (Y \times P_Y) - [(P_M \times X_M) + (P_H \times X_H) + (P_K \times X_K)] \tag{7}$$

Thus, other factors unchanged, $ATP^W(3)$ is (much) greater than $ATP^W(1)$ and $ATP^W(2)$.

Our theoretical critique of $ATP^W(3)$ is straightforward: We find case 3 to be an incorrect estimation of repayment capacity since it would require farm households not only to pay for water out of the residual returns to water but to possibly draw from the returns to all owned inputs. It would be correct only if the opportunity cost or price of all non-water-owned inputs were zero. Put another way, this approach to calculating ATP^W would – in addition to charging up to their total return to water – capture from farm households the return to some or all of their non-water-owned inputs. Economic theory suggests that if charges were set higher than $ATP^W(1)$, farm households would discontinue irrigated crop production.

2.5 Some additional issues

Space limits prevent more detailed analysis, but here we mention two other issues that arose in the Kyrgz Republic study. First, non-economists were eager to consider the value of livestock production and even of off-farm income in the assessment of farmers' repayment capacity. In our view, livestock operations are value-added enterprises that can exist largely independent of irrigation water, since feeds and forages for livestock production can, in principle, be purchased on the open market. The net return to water from feed and forage crops should already adequately reflect the contribution of irrigation water because forage crop outputs are valued using market prices, even though much of the production may be consumed on farm. Other on-farm enterprises, like clothing and handicrafts, also have little or no direct link to irrigation water. Off-farm employment, remittances, and social security payments are nonfarm sources of income that have no direct connection to water supply, and should not be tapped to help recover costs of supplying water. A second issue is how to price land in the calculations. Since the formulation is in annual terms, an annual land rental rate rather than a sales price is appropriate. A readily observed land rental rate is that for irrigated lands. However, market rental rates already would include the contribution of irrigation (as well as the on-farm capital cost of developing irrigated land (such as provision for land leveling, ditches, turnout structures, and drainage). The rental rate on nonirrigated land of comparable quality and productivity would be the appropriate measure of opportunity cost of land.

2.6 Further remarks

We should acknowledge that the brief description and the simplifying assumptions in the above analysis may make the process of estimating ATP appear to be a simpler task that it actually is. Space limitations only allow a mention of some of the challenges, mainly associated with the uncertainties of operating the long-lived investments characteristic of irrigation development or rehabilitation that should be recognized in both the WTP and ATP assessments. For example, assignment of opportunity costs of *noncontractual (owned) inputs* in the ATP and WTP analyses is complex and often controversial. Crop production and water application *technology* as well as real product and input *prices* can be expected to change over time. Over the past several decades, the improvement in crop production technology has led to a more or less countervailing reduction in inflation-adjusted commodity prices, so that static annual models assuming no change in technology, prices, or net income have shown better predictive value than attempts to independently predict price changes and technology and yield changes. (Whether this sort of simplification will continue to be adequate in a context of global climate change remains an open question.) Another consideration is the shorter-term income uncertainty due to *weather and markets*. (In 1982, Gittinger dealt with this issue by including a "risk-bearing" function in the definition of the production function and suggesting a charge of approximately 10% of net farm income to account for this concern.) In the case of farms at the tail end of an irrigation distribution system, water supply per unit area tends to be lower and, because of inadequate water availability for leaching salts, crop-damaging salinization of soils is greater. Thus, a given farm's *location on the system* has been shown to be an important determinant of water availability and crop production. If the charges analysis takes an average of head and tail incomes, the tailenders would be at a disadvantage.

2.7 Previous literature

As mentioned earlier, we found only a few efforts that directly addressed a conceptual framework for ATP. The World Bank has long emphasized enhancement of irrigation water supplies as an economic development tool in semi-arid and arid low-income countries throughout the world. Irrigation was perceived to yield favorable economic returns on donor agency and water user capital, as well as to enhance food security and employment in impoverished rural areas. The World Bank doubtless has evaluated and financed more irrigated land developments than has any other agency. Gittinger (1982) codified the Bank's procedures for evaluating agricultural projects and offered detailed descriptions of appropriate approaches to assessing cost recovery policies. In World Bank parlance (Gittinger, 1982; pp. 222–232, the calculation of what we call $WTP^{W}(2)$ is called the "economic analysis," and the calculation of approximately what we term $ATP^{W}(2)$ would be considered to be "financial analysis."

Moore (1999) studied a project of the U.S. Bureau of Reclamation in the Sacramento Valley in Northern California to measure farmers' ATP for irrigation water. He provides a theoretical definition of the term that we interpret as equivalent to our $ATP^{W}(1)$ and $WTP^{W}(1)$. That is, Moore's theoretical ATP is equal to estimated revenues less the cost of all non-water inputs, both contractual and noncontractual. With sophisticated econometric techniques, Moore fits a multi-output revenue function to 11 years of crop revenue, prices of major crops, areas of irrigated lands, and water delivery data from 19 irrigation districts that varied in water supply per acre. Comparing his measure of ATP with that of the water supply agency – the U.S. Bureau of Reclamation – Moore found that his method yielded a much larger estimate of ATP per unit area. Moore's inductive method is exceedingly elegant and, to its credit, relies on real-world observations. However, it is *ex post* in the sense that it must take data from an ongoing system, rather than studying a proposed future investment. Thus, in contrast to the typical deductive approach, and the one used here, it is not adapted for sensitivity analysis. These considerations suggest that the method would find its application mainly in research contexts.

3 Case study from the Kyrgyz Republic

3.1 Background: Description of study area

The Kyrgyz Republic became independent from the Soviet Union in 1991. Located in Central Asia, the largely mountainous country is bordered by Kazakhstan to the north, Uzbekistan to the west, Tajikistan to the southwest, and China to the east. Irrigated agriculture is centered in the higher short-season Chui and Talas river valleys in the north and east, and in the hot subtropical Ferghana Valley in the southwest. Major crops are feed and forages for livestock, plus grains such as wheat and barley.

Since the 1990s, external donors have been assisting the government in rehabilitating the extensive irrigation systems, and even in meeting the cost of routine operations and maintenance. Only a small fraction of the total cost of sustaining the country's irrigation systems has been met by charges assessed on farmers and other water users. It is increasingly recognized that the government needs to improve cost recovery from users, and that water user associations need to recover more from their members to sustain the infrastructure within their borders. With financial assistance from the Asian Development Bank, a study was initiated in 2006 to analyze the ability and willingness of farmers to pay for irrigation and drainage services[1] (Asian Development Bank, 2007).

The Sovhozhnyi branch canal in Chui Oblast was among the study areas selected as part of the study to assess the long-term sustainability of the irrigation system and, more specifically, the farmers' ATP and WTP for irrigation services. The Sovhozhnyi main canal services 4200 ha of level land that is managed by eight water user associations in the short-season Chui River valley. The study area is west and north of Bishkek, the country's capital.

3.2 Some preliminaries in the application of alternative models

As part of the study, representative farm budgets were constructed on the basis of a survey of farms within the Sovhozhnyi study area. The survey was stratified to ensure enough observations of farms by size (smaller, medium, and large) and location on the delivery canal (head, middle, and tail). The 2005 crop year, for which crop budget data were collected, was considered to be normal in terms of *yields* and *prices* of major commodities.

Regarding the use of *household labor*, since formal surveys enquiring about this input are often difficult for farmers to answer (and likely to underestimate it), a smaller number of more informal and in-depth farmer interviews were also conducted. *Management* reflects the time spent by the owner/manager in planning and other activities not captured in the time spent in field production and harvesting activities. This was valued at 5% of the total value of output. *Land* was valued at its opportunity cost, which was assumed to be the rental cost of nonirrigated, but otherwise similar, agricultural land. With very little capital at the farm level, *capital* was valued at zero.

3.3 Application of alternative models

To illustrate the application of the three alternative models, we calculated ATP and WTP values as residuals from a representative crop budget of 1 ha. As necessary, this can be scaled up to the respective values for an average farm or the entire study area.

The total value of output represents yield of all crops in the Sovhozhnyi study area (weighted by their proportion in the study area) multiplied by their output prices. It thus values all production whether actually sold or consumed on the farm or within the household. Given that crop prices in the Krygyz Republic are market-determined, with little government intervention, the total value of output is the same for the three models. Purchased input costs are also market-determined, and the same for all models. Costs for which the three conceptual models can lead to different values are those of labor, management, and land.

Table 1 shows estimates of WTP and ATP for water per hectare based on Model 1. This approach values all outputs and inputs at market prices. The estimated $\mathrm{ATP}^{W}(1)$ for irrigation services is US\$198 ha^{-1}, and equals the estimated WTP(1). All values are expressed in 2005 US\$, using the conversion rate of US\$1 = Som 40.

Table 1: Estimates for Model 1.

	$WTP^W(1)$ (US$ ha^{-1})	$ATP^W(1)$ (US$ ha^{-1})
Total value of production	381	381
Cost of production		
Seed, fertilizer, herbicides	21	21
Contracted services – labor	20	20
Contracted services – fuel	31	31
Transportation	20	20
Hired labor	17	17
Household labor	14	14
Management	19	19
Land cost	38	38
Miscellaneous	3	3
Total cost of production	183	183
Residual	198	198

Note: $WTP^W(1) = ATP^W(1)$.
Source: Based on data collected in Asian Development Bank TA 4405 "Study on Pricing Systems and Cost Recovery Mechanisms for Irrigation." Kyrgyz Republic. 2007.

Table 2 shows the respective estimates for Model 2. $WPT^W(2)$ is calculated with a shadow wage based on its social opportunity cost. Since there is a high level of unemployment and underemployment of labor in the Kyrgyz Republic that has existed since the economic collapse following the breakup of the Soviet Union, we chose as the opportunity cost for labor half of the market wage. No cost is attributed to management in $WTP^W(2)$. Gittinger (1982) argues that in the case of farmers who own their farms and supply most of both the labor and management skills, only wages as laborers should be valued, not management skills (though there are differences in opinion – see, e.g., Young (2005, Chapter 5). As a result, this estimate of the ATP for water is US$242 ha^{-1}, which is 22% greater than estimated by the model of ATP(1).

Estimates for Model 3 are shown in Table 3. $WTP^W(3)$ is assumed to be equal to $WTP^W(2)$. For $ATP^W(3)$, net farm income (essentially, return to farm-owned inputs) is the concept representing this version of the farmer's maximum ATP for water. The cost of farmer-owned inputs, including household labor, management, and land, are in this case valued at zero. This leads to an estimated ATP of US$268 ha^{-1}, or 36% higher than $ATP^W(1)$, and also higher than $WTP^W(3)$.

Table 2: Estimates for Model 2.

	$WTP^W(2)$ (US\$ ha^{-1})	$ATP^W(2)$ (US\$ ha^{-1})
Total value of production	381	381
Cost of production		
Seed, fertilizer, herbicides	21	21
Contracted services – labor	10	20
Contracted services – fuel	31	31
Transportation	20	20
Hired labor	9	17
Household labor	7	14
Management	0	19
Land cost	38	38
Miscellaneous	3	3
Total cost of production	139	183
Residual	242	198

Note: $WTP^W(2) > ATP^W(2)$ (with $ATP^W(2) = ATP^W(1)$).
Cost items that are shadow priced and thus differ from those in Model 1 are italicized.
Source: Based on data collected in Asian Development Bank TA 4405 "Study on Pricing Systems and Cost Recovery Mechanisms for Irrigation." Kyrgyz Republic. 2007.

Table 3: Estimates for Model 3.

	$WTP^W(3)$ (US\$ ha^{-1})	$ATP^W(3)$ (US\$ ha^{-1})
Total value of production	381	381
Cost of production		
Seed, fertilizer, herbicides	21	21
Contracted services – labor	10	20
Contracted services – fuel	31	31
Transportation	20	20
Hired labor	9	17
Household labor	7	0
Management	0	0

Land cost	38	0
Miscellaneous	3	3
Total cost of production	139	113
Residual	242	268

Note: $WTP^W(3) < ATP^W(3)$ (with $WTP^W(3) = WTP^W(2)$)
Cost items that are shadow priced and thus differ from those in Model 1 are italicized.
Source: Based on data collected in Asian Development Bank TA 4405 "Study on Pricing Systems and Cost Recovery Mechanisms for Irrigation." Kyrgyz Republic. 2007.

We find that estimated ATP in the study area using what we believe to be a conceptually correct formula (our $ATP^W(1)$) is considerably higher than the amounts farmers have actually been charged, and would thus justify some increase in farmer contributions; but it is rather less than local and donor officialdom (implicitly using our $ATP^W(3)$) had believed. Casual observations elsewhere in the world suggest that this situation is rather typical, so that ATP would tend to be lower than that assumed by non-specialists and non-economists.

4 Conclusions

ATP is an estimate of the maximum that farmers could afford to pay for irrigation services. Political realities suggest that it is unlikely that the government would wish to extract the entire ATP from smallholder farmers. Overestimating ATP would give irrigation districts, governments, and lending agencies an exaggerated perception of the amounts that farmers would be able to pay for services.

Comparison of the three models to estimating ATP, captured in Tables 1–3, shows that substantial differences in estimates are produced under the different models. The appropriate model for estimating ATP is Model 1. Valuing inputs and outputs at social opportunity costs may be an appropriate approach to take in a cost–benefit analysis, but carrying this approach over into ATP can lead to misleading high estimates. Similarly, returns to all farm-owned inputs is a valuable concept in short-run farm management studies, but using this approach to value long-run ATP for water alone also leads to excessively high estimates.

Thus, analysts calculating ATP need to give careful attention both to the theoretical model on which the calculations are based and to the economic and production conditions relevant to the calculations.

Endnote

[1]Asian Development Bank TA 4405 "Study on Pricing Systems and Cost Recovery Mechanisms for Irrigation." Kyrgyz Republic. 2007.

References

Cornish, G., Bosworth, B., Perry, C. & Burke, J. 2004. *Water Charging in Irrigated Agriculture: An Analysis of International Experience.* Water Reports 28, Food and Agriculture Organization of the United Nations, Rome.

Dinar, A. ed. 2000. *The Political Economy of Water Pricing Reforms.* Oxford University Press, New York.

Gittinger, J.P. 1982. *Economic Analysis of Agricultural Projects,* 2nd edition. Johns Hopkins University Press, Baltimore.

Kay, R. C., Edwards, W.M. & Duffy, P.A. 2007. *Farm Management, 6th edition.* McGraw Hill, Boston.

Molle, F. & Berkoff, J. eds. 2007. *Irrigation Water Pricing: The Gap between Theory and Practice.* CABI Wallingford, UK.

Moore, M.R. 1999. Estimating irrigators' ability to pay for reclamation water. *Land Economics*, 75(4) 562–578.

O'Mara, G. 1990. *Making Bank Irrigation Investments More Sustainable.* Agriculture and Rural Development Department, Working Paper WPS 420, World Bank, Washington, DC.

Sampath, R.K. 1992. Issues in irrigation pricing in developing countries. *World Development,* 20(7): 967–977.

Small, L. & Carruthers, I. 1991. *Farmer-Financed Irrigation: The Economics of Reform.* Cambridge University Press, Cambridge.

Tsur, Y., Roe, T., Doukkali, R. & Dinar, A. 2004. *Pricing Irrigation Water: Principles and Cases from Developing Countries.* Resources for the Future Press, Washington, DC.

Young, R.A. 2005. *Determining the Economic Value of Water: Concepts and Methods.* Resources for the Future, Washington, DC.

Chapter 11

Energy prices, biofuels, and irrigated agriculture

K. Schoengold[1], T. Sproul[2] & D. Zilberman[3,4]
[1]Department of Agricultural Economics, University of Nebraska-Lincoln, Lincoln, NE, USA
[2]Department of Agricultural and Resource Economics, University of California-Berkeley, Berkeley, CA, USA
[3]Department of Agricultural and Resource Economics and the Giannini Foundation of Agricultural Economics, University of California-Berkeley, Berkeley, CA, USA
[4]Research leading to this paper was supported by the Energy Bioscience Institute (EBI) and USDA ERS.

Abstract

Changing energy prices will alter water allocation and distribution. Water extraction and conveyance may become more costly and demand for hydroelectric power will grow. Increasing demand for hydroelectric power may reduce the price and increase supply of surface water. High energy prices may also lead to an expansion of water-intensive biofuel technology, resulting in a possible increase in the output price of food production. High energy prices may therefore have obvious implications for sustainable irrigation as well as food security. With a focus on surface water systems, the models considered in this chapter can be used to determine the effects of higher energy prices on water use, total agricultural production, and optimal water rates.

Keywords: Biofuels; Conveyance; Energy price; Surface water; Water price

1 Introduction

While the economic downturn of 2008 dampened energy prices, the era of the new millennium saw dramatically increasing energy prices reflecting increased global demand, production constraints, and concern about climate change. In addition, some nations have already passed renewable fuel mandates, leading to expansion of biofuels production. Irrigated agriculture is energy intensive, and biofuels compete with other crops for scarce water resources. High energy prices may therefore have significant implications for sustainable irrigation as well as food security. Thus, despite the current lull in energy prices, it is prudent to consider the impacts of increasing energy prices and expanded biofuels production on irrigated agriculture.

We model agricultural water systems with diversion, conveyance, and irrigation components and recognize the indirect impacts of higher energy prices via output prices and input costs. We analyze the impacts of these price changes on pumping decisions, investment in conveyance systems, and adoption of modern irrigation technologies. First, we present the results for a case when irrigation and conveyance technologies are fixed, and then shift to an analysis of how energy prices could impact technology choices. In each case, we focus on the impact of changes on three key variables: output levels, water use, and optimal water pricing.

2 Model of irrigation water demand

The analysis of agricultural water systems assumes they have several elements: diversion, conveyance, and irrigation. The first element is *diversion* from the source to a conveyance system. In the case of groundwater, diversion is pumping the water from an aquifer. We denote the cost of diversion, $C^D(X, p_e)$. These are the costs of building and maintaining water project infrastructure, where X is the volume of diverted water and p_e is the price of energy. The second element is *conveyance*. We assume that at the source, X units of water enter the conveyance system. In some systems, the water moves downhill from source to application, and in these cases the energy can be harnessed through hydropower production. In other cases, water is elevated from the source to the conveyance canal, a process that requires energy inputs. Let m denote the net conveyance energy requirement per unit of water. It is positive when water must be elevated and is negative when water moves downhill and hydropower can be harnessed. For simplicity, we assume that the elevation adjustment occurs at the source, and thus the cost of water transport is p_eXm. Conveyance efficiency, the fraction of water that is transmitted through the conveyance system, is denoted by η, with $0 < \eta \leq 1$. Let Z be the cost of lining the canal, so overall conveyance costs are $p_eXm + Z$.

The third element of the system is *irrigation*. Reflecting conveyance losses, the amount of applied water available for irrigation is $X^A = X\eta$. Not all the applied water is utilized in production, as some ends up as deep percolating water or runoff. Irrigation efficiency, the fraction of water actually utilized in production (which depends on irrigation technology), is denoted by γ, with $0 < \gamma \leq 1$. The effective water, which is the water actually consumed by the crop, is $X^E - X^A\gamma$. We assume that irrigation requires a fixed cost K and pressurization energy per unit of applied water, β.

Following Caswell and Zilberman (1986), we assume that output is a function of effective input. Since effective input is dependent on applied input and irrigation technology, output is a function of these two variables. The analysis implicitly assumes that irrigated agriculture is taking place on a given amount of land, but we do not introduce acreage explicitly. Other research by Chakravorty *et al.* (1995) shows the effect of various water policies and prices on the amount of land in production based on the distance from the water source. Let Y denote output, and the agricultural production function is $Y = f(X^E)$. The region of the production function where $f' > 0$, $f'' \leq 0$ is the economic region where profit-maximizing outcomes occur.

We consider a small region, for which output and energy prices are given. The price of output is denoted by p_o. Accordingly, the gross benefits of agricultural production, denoted by $GB(X) = p_o f(X\eta\gamma)$, are the revenues generated by selling the output produced by irrigation.

The extraction and application of water imposes environmental costs. Let these environmental costs be denoted $C^E(X)$. The marginal environmental costs of water are positive and increasing: $C_X^E > 0$, $C_{XX}^E \geq 0$.[1] We will use subscripts to denote derivatives throughout the paper, so $C_X^E = \partial C^E / \partial X$. We consider here the socially optimal choices, accounting for the production benefits from irrigation and the diversion, conveyance, and environmental costs of water use.

2.1 Optimal water use and output levels

Letting B denote the net social benefits of production,

$$B = \underset{X}{Max}\; p_o f(X\eta\gamma) - C^D(X, p_e) - p_e Xm - Z - K - p_e \beta X\eta - C^E(X) \quad (1)$$

The optimal water use, denoted by X^*, is determined by solving eqn (2):

$$p_o f_{X^E}(X^{E*})\eta\gamma - C_X^D(X^*, p_e) - p_e m - p_e \beta\eta - C_X^E(X^*) = 0 \quad (2)$$

where $X^{E*} = X^*\eta\gamma$ denotes optimal effective water and $p_o f_{X^E}(X^{E*})\eta\gamma = MGB(X^*)$ is the marginal gross benefits of diverted water used in irrigation, which are equal to the marginal product of diverted water in irrigation. The marginal productivity of effective water is $f_{X^E}(X^{E*})$, and the marginal productivity of diverted water is $f_{X^E}(X^{E*})\eta\gamma$, which is derived by multiplying by the input use efficiency and conveyance efficiency. These optimality conditions imply an optimal water price at the source, which is the sum of the marginal externality and extraction costs, and is denoted by

$$p^{\mathrm{wS}} = C_X^D(X^*, p_e) + C_X^E(X^*) \quad (3)$$

This is the optimal price that should be charged to water districts buying from a water project. This price is different from the optimal price of applied water at the field, p^{wA}:

$$p^{\mathrm{wA}} = \frac{C_X^D(X^*, p_e) + C_X^E(X^*) + p_e m}{\eta} = \frac{p^{\mathrm{wS}} + p_e m}{\eta} \quad (4)$$

This is the sum of the marginal diversion cost, the net marginal conveyance cost, and the marginal externality cost divided by the conveyance efficiency. p^{wA} is the optimal price that farmers should pay a water district for water delivered to the farm. Even in cases where conveyance does not create or use energy, the price of water at the farm will not be the same as the price at the source, because of conveyance losses.

Condition (2) suggests that optimal water use occurs when the marginal product of applied water in irrigation is equal to the price of applied water plus the marginal unit pressurization cost. Thus, eqn (2) implies

$$p_o f_{X^E}\left(X^{E*}\right)\gamma - p_e\beta - p^{wA} = 0 \tag{5}$$

The second-order condition that will assure that X^* is an optimum is

$$D = p_o f_{X^E X^E}\left(X^{E*}\right)(\eta\gamma)^2 - C_{XX}^D\left(X^*, p_e\right) - C_{XX}^E\left(X^*\right) < 0 \tag{6}$$

By our assumptions this condition holds.

3 Impact of changing energy prices

When examining the effect of the introduction of biofuels, we follow Zilberman *et al.* (2008) in assuming that the food and energy demand effects from increasing energy prices will increase output prices when biofuels are present. Thus, we introduce the production of biofuels in the model as $p_o'(p_e) > 0$. The optimality conditions in eqns (2), (3), and (5) enable analyzing the impacts of changes of energy prices on X^*, Y^*, p^{wS}, and p^{wA}.

Proposition 1:

(a) (1) $\frac{dX^*}{dp_e}$ and (2) $\frac{dY^*}{dp_e} \geq (\leq) 0 \; if \;\; p_o' f_{X^E}\eta\gamma - m \geq (\leq) C_{Xp_e}^D + \beta\eta$

(b) (1) $\frac{dp^{wS}}{dp_e} \geq (\leq) 0 \; if \; p_o' f_{X^E}\eta\gamma\sigma + C_{Xp_e}^D \geq (\leq) [m + \beta\eta]\sigma$

(2) $\frac{dp^{wA}}{dp_e} \geq (\leq) 0 \; if \; p_o' f_{X^E}\eta\gamma\sigma + C_{Xp_e}^D + m \geq (\leq) -\beta\eta\sigma$

where $\sigma = -\frac{C_{XX}^D + C_{XX}^E}{p_o f_{X^E X^E}(\eta\gamma)^2} > 0$, which is the relative concavity of cost versus benefit for total diverted water.

Proof: For part (a) totally differentiate eqn (2) to yield

$$\frac{dX^*}{dp_e} = -D^{-1}\left[p_o' f_{X^E}\eta\gamma - C_{Xp_e}^D - m - \beta\eta\right] \tag{7}$$

and note that $\frac{dY^*}{dp_e} = p_0 f_{X^E} \eta\gamma \frac{dX^*}{dp_e}$. For part (b), the differentiation of eqn (3) yields

$$\frac{dp^{wS}}{dp_e} = \frac{1}{\eta}\left[C^D_{Xp_e} + \left(C^D_{XX} + C^E_{XX}\right)\frac{dX^*}{dp_e}\right] \tag{8}$$

Introducing dX^*/dp_e^{NB} from eqn (7) and D from eqn (6) into eqn (8) and rearranging terms results in (b.1). Differentiation of eqn (4) yields

$$\frac{dp^{wA}}{dp_e} = \frac{1}{\eta}\left[\left(C^D_{Xp_e} + m\right) + \left(C^D_{XX} + C^E_{XX}\right)\frac{dX^*}{dp_e^{NB}}\right] \tag{9}$$

Introducing dX^*/dp_e and D into eqn (9) and rearranging terms yields (b.2).

Proposition 1(a) suggests the impact of higher energy prices on water use and output levels consists of four effects: the biofuel effect, $p_0' f_{X^E} \eta\gamma\sigma$, the diversion cost effect, $C^D_{Xp_e}$, the pressurization costs effect, $\beta\eta$, and the conveyance cost effect, m. Of these four effects, the biofuel effect is expected to increase water use and output levels (X^* and Y^*, respectively), the effects of diversion and pressurization costs are expected to decrease X^* and Y^*, and the effect of conveyance depends on whether it is a net creator of energy through hydropower. If the biofuel effect dominates, and the increase in energy price results in a greater marginal increase in earning than in costs, an increase in energy prices will lead to higher water use and production. When the increased costs resulting from higher energy price are greater than the increased revenues, higher energy prices lead to a reduction in water use and production. This result is shown graphically in Figure 1. Let

$$MC_j(X) = \frac{1}{\eta}\left(C^D_X\left(X, p_e^j\right) + C^D_X\left(X\right) + p_e^j m + p_e^j \beta\right) \text{ and}$$

$$MB_j(X) = p_0(p_e^j) f_X(X)\gamma\eta$$

where $j = 0,1$ denotes the old and new energy prices, respectively. If conveyance requires energy inputs, the marginal costs of water use will increase with higher energy prices. This is represented by a shift in the marginal cost curve from $MC_0(X)$ to $MC_1(X)$. At the same time, the biofuel effect increases the net benefits of water use. The relative magnitude of these two shifts is what determines the direction of water use and production levels. For example, if there is a shift from $MB_0(X)$ to $MB_1'(X)$, the effect of the shift in costs dominates the price effect, and total output and water use decrease. However, if there is a shift from $MB_0(X)$ to $MB_1''(X)$, the price effect dominates, leading to an increase in total output and water use.

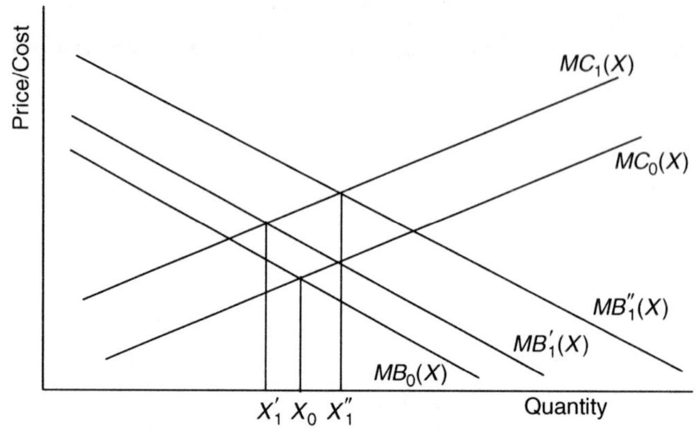

Figure 1: Relative impacts of changing output prices and input costs.

An increase in the price of energy *may* lead to increased water use and production even without the biofuel effect (i.e., $p_o^{\prime}=0$) in cases where conveyance generates hydroelectric power, and the marginal energy gain is greater than the marginal energy cost of diversion and pressurization, $-m \geq C_{Xp_e}^{D}(X^*, p_e)+\beta\eta$. In cases, when $p_o^{\prime}=0$ and conveyance consumes energy ($m > 0$), higher energy prices will lead to reduction in water use and output. These results suggest varying impacts of higher energy prices across locations. Higher energy prices may lead to reduced production in certain regions where conveyance consumes energy, such as elevated or mountain regions. But other regions, like valleys, where water sources are upstream and conveyance leads to hydroelectric power generation, will expand their irrigated agriculture or other water-dependent activities when energy prices rise.

3.1 Impacts of energy prices on optimal water prices

Proposition 1(b.1) suggests that with higher energy prices, the higher diversion costs will act to increase the price of water at the source, p^{wS}, and at the farm, p^{wA}, through an upward shift in the marginal cost curve. However, increased pressurization costs will have a negative effect on water prices due to a reduction in the quantity of water demanded. The coefficient σ is used to determine the combined effect of marginal changes in supply and demand on prices. The increase in output price because of biofuels will increase water prices through increased demand. If higher energy prices increase conveyance costs ($m > 0$), it will reduce the price at the source through lower demand, but increase the price to farmers (p^{wA}) through increased costs of supplying water. Figure 2 shows several possible scenarios for the impacts of increased energy prices on the optimal price of water at the source and at the field. Letting 0 and 1 denote the outcomes before and after an energy price change, we assume that $p_e^0 < p_e^1$. The relative magnitudes of X^0

and X^1 depend on the relative impacts of the biofuel and price effects, the extraction cost effect, and the pressurization cost effect.

If the diversion cost and biofuel price effects on water prices dominate the pressurization and conveyance cost effects, water prices at the farm gate and the source will both increase. This result is seen in Figure 2(a). If the conveyance cost effect is dominant, and it is positive (water needed to be lifted to the final users), higher energy price will result in lower water price at the source and higher prices to the final users. But if the dominant conveyance cost is negative (there are benefits from hydroelectric power), the water price at the source will increase due to higher demand, while the water price for the end users will decline; reflecting increased water availability and a downward shift of the supply curve. This final result is seen in Figure 2(c). Thus, increasing energy prices may increase water supply and reduce water prices for farmers who are downstream to a system that is a net generator of energy.

The effect of biofuel introduction: Proposition 1 suggests that the introduction of biofuels may alter the direction of the impacts of increased energy prices on agricultural production. To see this, the impacts of higher energy prices without biofuels are derived from below, from Proposition 1:

Corollary 1: Assuming no biofuel effect ($p_o' = 0$), and denoting outcomes without biofuel with a superscript NB, the following results hold.

$$\text{(a) (1)} \frac{dX^*}{dp_e^{NB}} \text{ and (2)} \frac{dY^*}{dp_e^{NB}} \leq (\geq) 0 \; if \; C_{Xp_e}^D + \beta\eta \geq (\leq) -m$$

$$\text{(b)} \quad \text{(1)} \; \frac{dp^{wS}}{dp_e^{NB}} \geq (\leq) 0 \; if \; C_{Xp_e}^D \geq (\leq) [m + \beta\eta]\sigma$$

$$\text{(2)} \; \frac{dp^{wA}}{dp_e^{NB}} \geq (\leq) 0 \; if \; C_{Xp_e}^D + m \geq (\leq) \beta\eta\sigma$$

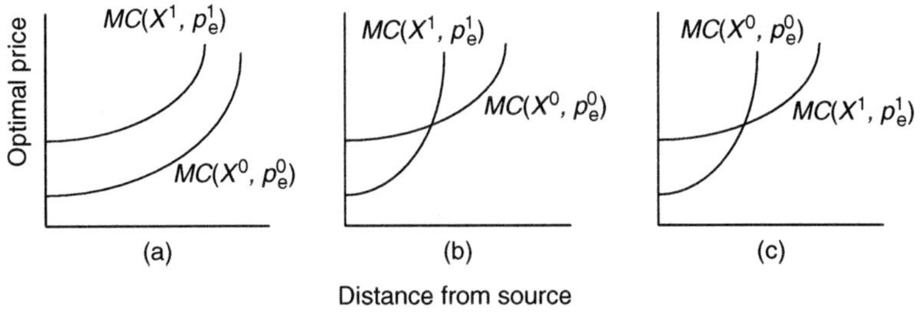

Figure 2: Impacts of higher energy prices on optimal water rates.

Remark: Comparison of Proposition 1(a) and Corollary 1(a) suggests that with biofuels, higher output prices may lead to increasing water use and production in situations when they would lead to reduction in irrigation agricultural activities without biofuels. One case is when conveyance consumes energy and

$$\frac{dX^*}{dp_e^{\mathrm{NB}}} \text{and} \frac{dY^*}{dp_e^{\mathrm{NB}}} < 0 .$$

However, this may still be the case when the derivatives are positive if the output price increase and the resulting biofuel effect are sufficiently large.

Comparison of Proposition 1(b) and Corollary 1(b) suggests that the introduction of biofuels tends to increase the price of water both at the source and on the farm. In some situations, large pressurization and conveyance effects may lead to reduction in the price of water at the source without biofuels, but the introduction of biofuels may actually lead to increase in prices if the biofuel effect exceeds the others. Furthermore, in situations where higher energy prices lead to reduction in conveyance costs ($m > 0$), they will reduce the on-farm water price without biofuels, but may increase the price *with* biofuels.

4 Choice of irrigation and conveyance technology

The management of the water system involves making several linked decisions, and the choice of irrigation technology. The conveyance losses can be affected by lining the canals. As presented in Section 2, l is an indicator of investment in lining, where $l = 0$ and $l = 1$ denote minimal and intensive lining, respectively. Conveyance efficiency, according to the conveyance investment choice, is denoted by η_l and $1 \geq \eta_1 \geq \eta_0 > 0$. The conveyance efficiency improvement of investing in more intensive lining is $\Delta\eta = \eta_1 - \eta_0 > 0$. Let Z_l be lining cost, and $Z_1 > Z_0 = 0$. Thus, overall conveyance costs are $p_e Xm + Z_l$.

We assume that producers determine both the water use level and the irrigation technology, i. We consider two technologies; the traditional is denoted by $i = 0$, and the modern by $i = 1$. We assume that $1 \geq \gamma_1 > \gamma_0 > 0$. We denote the irrigation efficiency improvement $\Delta\gamma = \gamma_1 - \gamma_0 > 0$. We assume that each technology requires a fixed cost K_i and pressurization energy per unit of applied water, β_i. For simplicity, assume that $K_0 = 0$, and thus the modern technology requires extra investment per acre. If the traditional technology is a gravitational one, and the modern one is drip or sprinkler, the modern technology is more energy intensive and $\Delta\beta = \beta_1 - \beta_0 > 0$. In cases where the transition is to a less energy-intensive technology (moving from sprinkler to a low-pressure irrigation), $\Delta\beta < 0$.

We consider here the socially optimal choices that are the results of an optimization that considers the production benefits from irrigation and the diversion, conveyance, and environmental costs of water use. The optimization problem is solved sequentially, where optimal water use is first determined for each discrete combination of irrigation technology and conveyance. Then, profits for each combination are compared and the combination that yields the highest profit level is chosen. Thus,

$$X_{i,l}^{*} = \underset{X}{\text{Max}}\ p_{\text{o}} f\left(X\eta_l\gamma_i\right) - C^D\left(X, p_{\text{e}}\right) - p_{\text{e}}Xm, \\ -Z_l - K_i - p_{\text{e}}\beta_i X\eta_l - C^E\left(X\right) \forall \{i,l\} \tag{10}$$

$$\{i^*, l^*\} = \underset{i,l}{\text{Argmax}}\ \{p_{\text{o}} f\left(X_{i,l}^{*}\eta_l\gamma_i\right) - C^D\left(X_{i,l}^{*}, p_{\text{e}}\right) - p_{\text{e}}X_{i,l}^{*}m, \\ -Z_l - K_i - p_{\text{e}}\beta_i X_{i,l}^{*}\eta_l - C^E\left(X_{i,l}^{*}\right)\} \tag{11}$$

Thus far we have analyzed the impacts of changes in energy when the irrigation and conveyance technologies are fixed, but in the longer run, higher energy prices may lead to investments in improved irrigation or conveyance. To analyze adoption behavior, it is necessary to derive the impacts of adoption on water use, output, and profits so the costs and benefits of adoption can be assessed. The impacts of adoption on water use and output are derived in this section. To simplify the analysis, we define two parameters of the production function, the output elasticity of applied water, ψ, and the elasticity of marginal productivity, ϕ, where both parameters are functions of effective water:

$$\psi\left(X_{i,l}^{E*}\right) = \frac{f_{X^E}\left(X_{i,l}^{E*}\right)X_{i,l}^{E*}}{f\left(X_{i,l}^{E*}\right)} > 0 \tag{12a}$$

$$\phi\left(X_{i,l}^{E*}\right) = -\frac{f_{X^E X^E}\left(X_{i,l}^{E*}\right)X_{i,l}^{E*}}{f_{X^E}\left(X_{i,l}^{E*}\right)} > 0 \tag{12b}$$

In the case of a Cobb–Douglas production function, both elasticities are constant and under concavity both ψ, $\phi \leq 1$. But as Caswell and Zilberman (1986) demonstrate, output response to water does not fit the Cobb–Douglas specification well. With irrigation, output peaks at a certain level, and overirrigation may lead to loss of yield. Studies like Hexem and Heady (1978) show that the quadratic production function fits data better than Cobb–Douglas, and thus it is more realistic to assume that the elasticities vary with effective input use, and analyze their properties using the more general assumptions proposed by Knight (1967). The classical Knightian production function has three phases including the one with decreasing marginal productivity. At the point where output is reaching its peak (D in Figure 3), $\psi = 0$ and the denominator of ϕ is zero; thus $\phi = \infty$. When the marginal productivity of effective water is peaking (A in Figure 3), $\phi = 0$ $\phi = 0$, and thus ϕ is increasing as effective water and output increase. It is useful to divide the economic region between B and D, where the marginal productivity

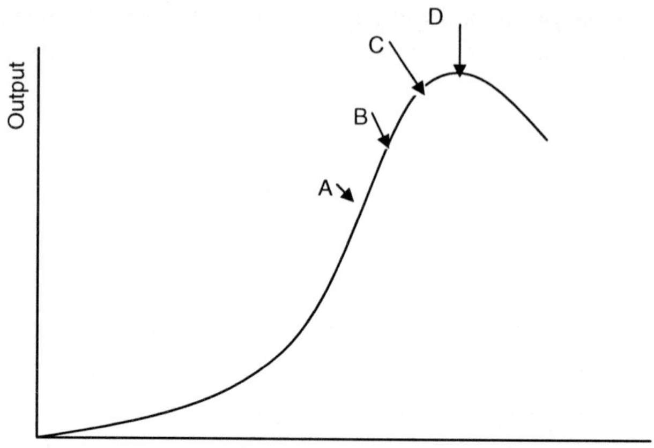

Figure 3: Output as a function of effective water.

of effective water is below its average productivity to two subregions. The first is between B and C in Figure 3, where irrigation is relatively sparse (relatively low effective water use and output), the marginal productivity of X^E (effective water) is declining only slightly, and the elasticity of marginal productivity (hereafter, EMP) is low, $\phi < 1$. The second region is where irrigation is plentiful, between C and D. In this region, output is close to its peak and the EMP is high ($\phi > 1$). As we will be shown later, the impact of adoption on water use and output will depend on whether water is sparse or plentiful.

4.1 Impacts of technology adoption on water use and output

The discrete choice of conveyance and irrigation technologies is chosen once $X^*_{i,l}$ is determined for all the possible combinations. We analyze the difference in production level, diverted water, and applied water under different technologies and determine how they are affected by changes of key parameters. Let $\Delta Y^*_l = Y^*_{1,l} - Y^*_{0,l}$ be the change in output associated with adoption of modern irrigation technology (i changes) with *fixed* conveyance configuration l. Similarly, $\Delta X^*_l = X^*_{1,l} - X^*_{0,l}$ is the diverted water differential between modern and traditional irrigation *with conveyance fixed.* Since our solution to the optimization problem is unique by concavity in eqns (4) and (5), $X^*_{i,l}$ can be written as $X^*(\gamma_i,\beta_i,\eta_l)$, that is, X^* evaluated at the point $(\gamma_i,\beta_i,\eta_l)$. Since the parameters γ,β,η are real-valued continuous variables within [0,1], where $i = 0,1$ and $l = 0,1$ simply denote fixed values of each parameter to identify the change due to technological adoption, and since X^* is a continuous function over that interval, we can use the Mean Value Theorem to express the effect of the discrete choice of irrigation technology i on $X^*_{i,l}$, in terms of partial derivatives:

$$\begin{aligned}\Delta X_l^* &= X^*(\gamma_1,\beta_1,\eta_l) - X^*(\gamma_0,\beta_0,\eta_l) \\ &= \frac{dX^*(\gamma_\varepsilon,\beta_\varepsilon,\eta_l)}{d\gamma}(\gamma_1-\gamma_0) + \frac{dX^*(\gamma_\varepsilon,\beta_\varepsilon,\eta_l)}{d\beta}(\beta_1-\beta_0)\end{aligned} \tag{13}$$

where $\gamma_\varepsilon \subset (\gamma_0,\gamma_1)$ and $\beta_\varepsilon \subset (\beta_0,\gamma_1)$ or $\beta_\varepsilon \subset (\beta_1,\beta_0)$, whichever is nonempty depending on the sign of $\Delta\beta$. From here forward we will denote $X^*_{\varepsilon,l} = X^*(\gamma_\varepsilon,\beta_\varepsilon,\eta_l)$. Similarly, we derive the discrete conveyance choice, holding irrigation technology fixed:

$$\Delta X_i^* = \frac{dX^*(\gamma_i,\beta_i,\eta_\varepsilon)}{d\eta}(\eta_1-\eta_0) \tag{14}$$

where $\eta_\varepsilon \subset (\eta_0,\eta_1)$ and $X^*_{i,\varepsilon}$ will denote $X^*(\gamma_i,\beta_i,\eta_\varepsilon)$.

To analyze the impacts of properties of ΔX_i^*, ΔX_η^* and ΔY_i^* and ΔY_η^* we need to derive the derivatives of X_i^* and Y_i^* with respect to β, γ, and η. The impact of changes in the parameters on $X^*_{i,l}$ and $Y^*_{i,l}$ are derived from eqn (2).

$$\begin{aligned}&(a)\ \frac{dX^*_{i,l}}{d\gamma_i} = \frac{X^*_{i,l}\chi[1-\phi]}{\gamma_i\phi} > 0 \ iff\ \phi < 1 \\ &(b)\ \frac{dY^*_{i,l}}{d\gamma_i} = \psi\frac{Y^*_{i,l}}{\gamma_i}\left[\frac{\gamma_i dX^*_{i,l}}{X^*_{i,l}d\gamma_i}+1\right] = \psi\frac{Y^*_{i,l}}{\gamma_i}\left[\frac{\chi[1-\phi]}{\phi}+1\right] > 0\end{aligned} \tag{15}$$

where $\chi = \frac{p_0 f_{X^E X^E}(\eta_l\gamma_i)^2}{D} \le 1$ is an indicator of the contribution of marginal productivity to the concavity of profit with respect to water.[2] Equations (15a) and (15b) suggest that an increase in input use efficiency (irrigation efficiency, in this case) will increase water demand and agricultural output in cases when water is sparse and output is far from maximum yield levels (EMP<1). However, both water use and output will decline with increased efficiency when the marginal productivity of water is low and EMP>1. This is more likely to occur in regions where the cost of water is low and there are few limitations on irrigation water usage. When improved irrigation efficiency increases water use, it will also increase output, but there are cases where output can increase even as water use declines. We find a similar result for improvements in conveyance efficiency in eqn (16):

$$\begin{aligned}&(a)\ \frac{dX^*_{i,l}}{d\eta_l} = \frac{X^*_{i,l}\chi}{\eta_l\phi}(1-\phi-s_\beta) > (\le) 0 \ if\ \phi < (\ge) 1 - s_\beta \\ &(b)\ \frac{dY^*_{i,l}}{d\eta_l} = \psi\frac{Y^*_{i,l}}{\eta_l}\left[\frac{\eta_l dX^*_{i,l}}{X^*_{i,l}d\eta_l}+1\right] = \psi\frac{Y^*_{i,l}}{\eta_l}\left[\frac{\chi(1-\phi-s_\beta)}{\phi}+1\right] > 0\end{aligned} \tag{16}$$

[3]

where $s_{\beta} = \dfrac{p_e \beta \eta_l}{p_o f_{X^E} \gamma_i \eta_l} \leq 1$ is the marginal pressurization cost of water relative to the value of marginal product of water in production.

Increasing conveyance efficiency has two effects on diverted water – the first is *de facto* increasing the use efficiency of diverted water, and thus has an effect that depends on the EMP, and the second is increasing the amount applied, leading to extra pressurization costs that will tend to decrease demand, especially with high-pressure sprinkler irrigation. As with irrigation efficiency, if conveyance efficiency increases water use then it will also increase output, but there will be cases where output increases as water use declines. Equation (16) examines the effect of changes in energy intensity of pressurization on output and water use:

$$\begin{aligned} &(a)\ \frac{dX_{i,l}^*}{d\beta_i} = -\frac{X_{i,l}^* \chi}{\beta_i \phi} s_{\beta} < 0 \\ &(b)\ \frac{dY_{i,l}^*}{d\beta_i} = -\psi \frac{Y_{i,l}^* \chi s_{\beta}}{\beta_i \phi} < 0 \end{aligned} \tag{17}$$

Equations (17a) and (17b) show that increased energy requirements of pressurization would reduce water use and output since the energy must come at some cost. The opposite is also true; lower energy requirements will increase optimal diversion. Note that this effect is proportional to the price of energy. The above formulas enable us to assess the impacts of adoption of conservation and improved conveyance on water use and output.

Proposition 2: (a.1) $\Delta X_l^* \leq (\geq) 0$ when $\phi \geq (\leq) 1 - s_{\beta} \dfrac{\Delta\beta}{\beta_0} \bigg/ \dfrac{\Delta\gamma}{\gamma_0}$

(a.2) $\Delta X_i^* \leq (\geq) 0$ when $\phi \geq (\leq) 1 - s_{\beta}$

(b.1) $\Delta Y_l^* \geq (\leq) 0$ when $\left(1 - \dfrac{\chi[1-\phi]}{\phi}\right) \dfrac{\Delta\gamma}{\gamma_0} - \dfrac{\chi s_{\beta}}{\phi} \dfrac{\Delta\beta}{\beta_0} \geq (\leq) 0$

(b.2) $\Delta Y_i^* \geq 0$

Proof: Inserting $\dfrac{dX_{i,l}^*}{d\gamma_i}$ and $\dfrac{dX_{i,l}^*}{d\beta_i}$ from eqns (14.a) and (16.a) into eqn (12) yields

$$\Delta X_l^* = \frac{X_{i,l}^* \chi}{\phi} \left[[1-\phi] \frac{\Delta\gamma}{\gamma_i} - s_{\beta} \frac{\Delta\beta}{\beta_0} \right] \tag{18}$$

which implies (a.1). Similarly inserting $\dfrac{dX_{i,l}^*}{d\eta_l}$ from eqn (16.a) into eqn (14) yields

$$\Delta X_i^* = \frac{X_{i,l}^* \chi}{\eta_l \phi} \left(1 - \phi - s_{\beta}\right) \Delta\eta \tag{19}$$

which after substitution for ψ yield (a.2). To derive (b.1) note that

$$\Delta Y_l^* = \frac{dY_{\varepsilon,l}^*}{dX}\Delta X_l^* + \frac{dY_{\varepsilon,l}^*}{d\gamma}\Delta\gamma = f_{X^E}\gamma_i\eta_l\Delta X_l^* + f_{X^E}\eta_l X_l^*\Delta\gamma \tag{20}$$

Since $f_{X^E}\gamma_i\eta_l = \psi Y / X$ and $f_{X^E}X\eta_l = \psi Y / \gamma_i$, eqn (19) becomes

$$\Delta Y_l^* = \psi\left(X_{\varepsilon,l}^{E*}\right)Y_{\varepsilon,l}^*\left[\frac{\Delta X_l^*}{X_{ol}^*} + \frac{\Delta\gamma}{\gamma_0}\right] \tag{21}$$

Dividing and multiplying parts of eqn (17) by $p_o f_{X^E X^E}(\eta_l\gamma_i)^2$, $\Delta X_l^* / X_{ol}^*$ can be rewritten as:

$$\frac{\Delta X_l^*}{X_{ol}^*} = \chi\frac{[1-\phi]}{\phi}\frac{\Delta\gamma}{\gamma} + \frac{p_e\Delta\beta\eta_l}{D} \tag{22}$$

Inserting eqn (22) into eqn (21) yields

$$\Delta Y_l^* = \psi\left(X_{\varepsilon,l}^{E*}\right)Y_{\varepsilon,l}^*\left[\left(1+\chi\frac{[1-\phi]}{\phi}\right)\frac{\Delta\gamma}{\gamma} + \frac{p_e\Delta\beta\eta_l}{D}\right] \tag{23}$$

which implies (b.1). To derive (b.2), note that

$$\Delta Y_i^* = \psi\left(X_{i,\varepsilon}^{E*}\right)Y_{i,\varepsilon}^*\left[\frac{\Delta X_i^*}{X_{i0}^*} + \frac{\Delta\eta}{\eta_0}\right] \tag{24}$$

By manipulating eqn (19), $\Delta X_i^* / X_{i0}^*$ can be rewritten as follows:

$$\frac{\Delta X_i^*}{X_{i0}^*} = \frac{\chi}{\phi}\left(1-\phi-s_\beta\right)\frac{\Delta\eta}{\eta_0} \tag{25}$$

Inserting eqn (25) into eqn (24) yields

$$\Delta Y_i^* = \psi\left(X_{i,\varepsilon}^{E*}\right)Y_{i,\varepsilon}^*\left[1+\frac{\chi}{\phi}\left(1-\phi-s_\beta\right)\right]\frac{\Delta\eta}{\eta_0} \tag{26}$$

which results in (b.2).

Remark: The change in water use after adoption of irrigation technology depends on the effects of irrigation efficiency gain and change in pressurization cost. The pressurization cost has a negative (positive) effect if $\Delta\beta > 0$ ($\Delta\beta < 0$), and the efficiency gain has a positive or negative effect depending on the EMP. From Proposition 2a.1, for the case with $\Delta\beta = 0$ $\Delta\beta = 0$, when the initial irrigation is sparse and the EMP<1, the efficiency gain associated with adoption will lead to increased water use. But, when the initial irrigation is plentiful and the EMP>1, initial output is close to the peak and the main gain from increased irrigation efficiency is water conservation, so adoption will lead to reduced water use.

Thus, an increase in pressurization needs due to irrigation technology adoption will reduce the critical level of the EMP (below which adoption will increase water use) to less than unity. A greater increase in pressurization costs relative to the value of the productivity gain reduces the critical level of the EMP because of the adoption effect. In cases where users of sprinkler irrigation switch to drip or other low-pressure irrigation, the pressurization cost effect is positive, so these users will increase water use relative to users adopting technologies requiring additional pressurization. The pressurization cost effect will increase under higher energy prices, making water conservation a more likely result. Higher output prices will have the opposite effect, since they increase the value of any productivity gain from increased water use.

Similarly, $Y^*_{i,l}$ can be written as $Y^*\left(\gamma_i, \beta_i, \eta_l\right)$, so

$$\Delta Y_l^* = \frac{dY^*_{\varepsilon,l}}{dX}\left[\frac{dX^*_{\varepsilon,l}}{d\gamma}\Delta\gamma + \frac{dX^*_{\varepsilon,l}}{d\beta}\Delta\beta\right] + \frac{dY^*_{\varepsilon,l}}{d\gamma}\Delta\gamma \qquad (27)^\dagger$$

$$\Delta Y_i^* = \frac{dY^*_{i,\varepsilon}}{dX}\left[\frac{dX^*_{i,\varepsilon}}{d\eta}\Delta\eta\right] + \frac{dY^*_{i,\varepsilon}}{d\eta}\Delta\eta \qquad (28)$$

We present ΔY_l^* in unexpanded form here for clarity; the full details and sign derivation can be found in the mathematical appendix. Equation (16) shows that the change in output as a result of adopting improved irrigation technology is the sum of two effects. The final term $\left(dY^*_{\varepsilon,l}/d\gamma\right)\Delta\gamma$ represents the direct effect of improved irrigation efficiency on output. The first term, on the other hand, represents the indirect effect of changes in water use on output, which depend (as demonstrated previously) on the net effect of productivity gains from increased irrigation efficiency and on increased costs due to energy requirements of pressurization.

In cases where the pressurization effect is positive or zero, such as when moving from sprinkler irrigation to drip irrigation or from drip irrigation to a lower pressure type, output will increase unambiguously when more efficient irrigation technology is adopted. However, when moving from a more primitive furrow irrigation system to drip or sprinkler, it is possible that increased pressurization energy costs exceed the value of productivity gained due to more efficient irrigation. Even in this case, where water use decreases after adoption, it is still possible that the direct effect dominates and output grows, which follows directly since $\Delta Y_l^* > 0$ under broader conditions than $\Delta X_l^* > 0$. Thus, an increase in water use due to switching to more efficient irrigation technology implies an increase in output as well, but the converse is not necessarily true. Higher output prices will make increasing water use under efficient irrigation more likely, while higher pressurization requirements and higher energy prices will have the opposite effect.

Only when initial water use is low (and EMP is consequently high) is it possible for the indirect effect to dominate and output to decrease. In other words, output will only decrease after adoption of conveyance improvement in water-conserving regions using improved conveyance efficiency as a water-saving tool. There will also be regions where the indirect effect is small, but negative, so that water is conserved even as output increases.

In some cases, there may be negative environmental impacts of adopting precision irrigation or shifting to improved conveyance technology. This could occur because these technologies increase the ratio of effective water to applied water, reducing the amount of runoff. In regions where the runoff from applied water is used to recharge aquifers or increase streamflow, more efficient irrigation technology reduces the amount of water that provides these environmental benefits. These results will not occur in cases where irrigators only have rights to their consumptive use. However, in most cases water rights are distributed based on applied water use, not consumptive water use. Therefore, these results may occur with the adoption of precision technology.

5 Impact of changing energy prices on technology choice

To analyze the impact of changes in the key parameters on the adoption of modern technologies or introduction of efficient conveyance, we will analyze the impact of parametric changes on the difference in the net benefits between alternative discrete solutions. Let $\Delta B_l = B_{1,l} - B_{0,l}$ denote the net gain from introduction of modern irrigation technology holding conveyance investment fixed, and let $\Delta B_i = B_{i,1} - B_{i,0}$ be the change in net benefits due to switching to improved canal lining for efficient conveyance, while holding irrigation technology i fixed. Technology adoption occurs when it provides a net gain in social benefit, that is, when $\Delta B_l > 0$ and/or when $\Delta B_i > 0$:

$$\Delta B_l = p_o \Delta Y_l^* - \Delta X_l^* \left(C_X^D \left(X_{\varepsilon,l}^*, p_e \right) + p_e m + C_X^E \left(X_{\varepsilon,l}^* \right) \right) - K_1 - p_e \eta_l \Delta X_l^{B*} \quad (29)$$

$$\Delta B_i = p_o \Delta Y_i^* - \Delta X_i^* \left(C_X^D \left(X_{i,\varepsilon}^*, p_e \right) + p_e m + C_X^E \left(X_{i,\varepsilon}^* \right) \right) - Z_1 - p_e \beta_i \Delta X_i^{A*} \quad (30)$$

where $X_{i,l}^{A*} = \eta_i X_{i,l}^*$ and $X_{i,l}^{B*} = \beta_i X_{i,l}^*$, following our previous notation for $X_{i,l}^{E*}$, and $\Delta X_i^{A*} = X_{i,1}^{A*} - X_{i,0}^{A*}$ and $\Delta X_l^{B*} = X_{1,l}^{B*} - X_{0,l}^{B*}$. Thus, if water use and output increase, the change in social benefit is positive if the gain from added output value is greater than the increased conveyance, diversion, environmental, and pressurization costs of water plus the fixed cost of the technological improvement. Output and water use may both decrease, in which case adoption *would* occur if the savings of water-associated costs outweigh the lost output value and fixed technology cost. Thus, adoption is most likely when output increases and water use decreases, since the cost of technology is offset by the sum of added output value and saved water-associated costs. In the appendix, we derive reduced forms of eqns (29) and (30):

$$\Delta B_l = X^*_{\varepsilon,l}\eta_l\left[p_o\Delta\gamma - p_e\Delta\beta\right] - K_1 \quad (31)^\dagger$$

$$\Delta B_i = X^*_{i,\varepsilon}\Delta\eta\left[p_o\gamma_i - p_e\beta_i\right] - Z_1 \quad (32)^\dagger$$

In each equation, the first term represents the total net value of efficiency gained from technology adoption and the second term is the fixed cost thereof. Specifically, the bracketed term in eqn (31) is the net unit value of efficiency gained for applied water, defined by the unit irrigation efficiency output gain less the increased unit energy cost of pressurization. In eqn (32), $X^*_{i,\varepsilon}\Delta\eta$ is the gain in applied water due to conveyance improvement and the bracketed term weights it by the net unit output value of fixed irrigation technology i. Thus, if the total net value of efficiency gained by adopting a modern technology is greater than the capital cost of the technology, then the technology will be adopted. Water-intensive regions where increased pressurization costs dominate the value of output gained, and water-conserving regions where water-savings come at too high of a cost (in terms of lost production value) will not adopt.

Proposition 3: Biofuels increase the benefits of adopting water conservation technologies under higher energy prices.

The likelihood of adoption of irrigation and conveyance technologies is affected by increasing energy prices depending on the sign of $d\Delta B_l/dp_e$ and $d\Delta B_i/dp_e$. We begin by examining the effects of energy price changes without biofuels production, where *NB* denotes "no biofuels."

$$\begin{aligned}\frac{d\Delta B_l}{dp_e^{NB}} &= \eta_l\frac{dX^*_{\varepsilon,l}}{dp_e}\left(p_o\Delta\gamma - p_e\Delta\beta\right) - \eta_l X^*_{\varepsilon,l}\Delta\beta \\ &\geq(\leq)0 \text{ if } \frac{dX^*_{\varepsilon,l}}{dp_e}\left(p_o\Delta\gamma - p_e\Delta\beta\right) - X^*_{\varepsilon,l}\Delta\beta \geq(\leq)0\end{aligned} \quad (33)$$

$$\begin{aligned}\frac{d\Delta B_i}{dp_e^{NB}} &= \frac{dX^*_{i,\varepsilon}}{dp_e}\Delta\eta\left(p_o\gamma_i - p_e\beta_i\right) - X^*_{i,\varepsilon}\Delta\eta\beta_i \\ &\geq(\leq)\,0 \text{ if } \frac{dX^*_{i,\varepsilon}}{dp_e}\left(p_o\gamma_i - p_e\beta_i\right) - X^*_{i,\varepsilon}\beta_i \geq(\leq)\,0\end{aligned} \quad (34)$$

Thus, the impact of higher energy prices on the benefits from adopting conservation technologies depends on the competing effects of the indirect and direct margins. In both equations above, the first term represents the indirect margin – the marginal effect of energy prices on optimal water use and the corresponding benefits of production value (or increased production value) versus the costs of pressurization (or increased costs thereof). The second term represents the direct margin – the extra pressurization costs for all water used due to higher energy prices. These results suggest that the benefits of adopting conservation technologies are decreased by higher energy prices unless the intensive margin benefits dominate the direct margin costs.

To consider the effect of energy price changes on technology adoption when biofuels are produced, we reintroduce the assumption that $p_o'(p_e) > 0$, and thus there is a possibility that demand effects are strong enough (even without significant hydropower gains) to make $dX^*/dp_e > 0$. As in Proposition 1, BF denotes the presence of biofuels production. The relationship between technology adoption and energy prices with biofuels is

$$\frac{d\Delta B_l}{dp_e^{BF}} = \frac{dX^*_{\varepsilon,l}}{dp_e}\eta_l\left[p_o\Delta\gamma - p_e\Delta\beta\right] + X^*_{\varepsilon,l}\eta_l\left[p_o'\Delta\gamma - \Delta\beta\right] \tag{35}$$

$$\frac{d\Delta B_i}{dp_e^{BF}} = \frac{dX^*_{i,\varepsilon}}{dp_e}\Delta\eta\left[p_o\gamma_i - p_e\beta_i\right] + X^*_{i,\varepsilon}\Delta\eta\left[p_o'\gamma_i - \beta_i\right] \tag{36}$$

The key difference in the measure of net benefits (due to rising energy prices) when biofuels are produced is the demand effect of biofuels. This demand makes it possible for the direct margin to have a positive impact on benefits when energy prices change. We compare eqn (33) with eqn (35) and eqn (34) with eqn (36) to show that biofuels increase the benefits of adopting water conservation technologies under higher energy prices:

$$\frac{d\Delta B_l}{dp_e^{BF}} - \frac{d\Delta B_l}{dp_e^{NB}} = p_o'(p_e)X^*_{\varepsilon,l}\eta_l\Delta\gamma > 0 \tag{37}$$

$$\frac{d\Delta B_i}{dp_e^{BF}} - \frac{d\Delta B_i}{dp_e^{NB}} = p_o'(p_e)X^*_{i,\varepsilon}\Delta\eta\gamma_i > 0 \tag{38}$$

The amount of increased benefits from the presence of biofuels production depends on the demand effect, the level of optimal water use, and the efficiency gains from the modern technology being adopted.

6 Conclusion

In this chapter, we have shown that a change in energy prices has several possible effects on irrigated agriculture. Energy is an input into agricultural production needed for the extraction, conveyance, and pressurization of irrigation water. If energy comes entirely from fossil fuels, the directional impacts of an increase in the price of energy are clear. The total amount of water used in production will decrease, and the level of output will also decrease. This outcome would be beneficial for sustainable irrigation, as total irrigation water is expected to decrease. However, the impacts on food security are a potential concern. If there is a reduction in total output and irrigation water use, food price increases may have an adverse effect on food security, especially for the poorest groups in society. The optimal water price at the source is likely to decrease due to a demand effect, but the optimal price of delivered water may increase if water conveyance requires energy inputs. However, there are several caveats to this basic story.

If higher energy prices are correlated with an expansion in the production of biofuels, the impacts on irrigated agriculture are less certain. An expansion in biofuel production is expected to increase output prices, which increases optimal input quantities of irrigation water and optimal yield levels. This effect, which we call the *biofuel effect*, results in producers using their land more intensely after biofuels are introduced. In some cases, this effect may dominate the *input price effect* and lead to a net increase in irrigation water inputs and output. The impact on food security is unclear, as agricultural producers benefit from higher output prices, and thus have greater food security, as long as the income effects from higher output prices outweigh the higher prices of food purchases. For the urban poor, higher energy prices compounded by the expansion of biofuels will clearly reduce their level of food security. This is because food prices are expected to increase both because of the higher price of energy as well as the diversion of grain stock into biofuel production. In addition, it is often the case that surface water systems deliver water to irrigators and also create hydropower through conveyance, resulting in a *conveyance effect* of higher energy prices. If this is the case, it is possible that higher energy prices will increase the net benefit of delivered water for irrigation, and thus may lead to an expansion of water consumption and output even if the biofuel effect is zero. In both cases, the expansion of irrigation water usage may result in unsustainable irrigation practices. Because of the conveyance effect, the impact of higher energy prices on optimal water rates may differ depending on whether water is priced at the source or at the field. If conveyance creates energy through hydropower, optimal water rates will be relatively higher at the source than after delivery, while the opposite is true if conveyance requires net energy inputs.

In addition to the net effect of higher energy prices on a fixed irrigation system, we also consider the possibility that producers can adopt improved irrigation or conveyance systems if energy prices increase. First we consider the impact of adopting improved irrigation or conveyance technology on water use and output levels. In cases where the marginal product of effective water (EMP) is high, it is likely that improved irrigation or conveyance efficiency will increase the use of water and output levels. In other cases where EMP is low (i.e., where output levels are close to maximum yield levels), a shift to improved irrigation or conveyance systems is likely to decrease total water use, while output levels might increase or decrease depending on the strength of the various effects. Higher energy prices, especially when they are combined with an expansion of biofuels, will increase the likelihood that producers adopt both precision irrigation systems and improved conveyance infrastructure. Such technologies may help with sustainable irrigation goals for the long run. The long-run sustainability of agriculture in an era of increasing energy scarcity and climate change depends on continued improvement in productivity that will reduce the need to expand utilization of water resources and increase the price of food. This implies the need for public sector commitment to support of agricultural research and to a regulatory system that allows innovations to be adopted in an efficient and environmentally sound manner.

Endnotes

†Additional details on the derivation included in the mathematical appendix.

[1]We do not have a strong a priori knowledge on how higher energy prices affect environmental costs, so they depend on diverted water here.

[2]From total differentiation of the FOC one obtains $\frac{dX^*_{i,l}}{d\gamma_i} = -\frac{p_o \eta_l f_{X^E}\left(X^{E*}_{i,l}\right)[1-\phi]}{D}$ and by substituting for χ and ϕ obtains 14(a).

[3]$\frac{dY^*_{i,l}}{d\eta_l} > 0$ since $\frac{\chi\left(1-\phi-s_\beta\right)}{\phi} + 1 = \frac{\phi(1-\chi)+\chi\left(1-s_\beta\right)}{\phi} > 0$

References

Caswell M. & Zilberman, D. 1986. The effects of well depth and land quality on the choice of irrigation technology. *American Journal of Agricultural Economics*, 68: 798–811.

Chakravorty, U., Hochman, E. & Zilberman, D. 1995. A spatial model of optimal water conveyance. *Journal of Environmental Economics and Management*, 29: 25–41.

Hexem, R.W. & Heady, E.O. 1978. *Water Production Functions for Irrigated Agriculture*. The Iowa State University Press, Ames, IA, USA.

Knight, F.H. 1967. *The Economic Organization: With an Article, Notes on Cost and Utility*. AM Kelley, New York.

Zilberman, D.T., Sproul, D., Rajagopal, S., Sexton, S. & Hellegers, P. 2008. Rising energy prices and the economics of water in agriculture. *Water Policy*, 10: 11–21.

Mathematical appendix

More detail and sign derivation for eqn (27)

$$\Delta Y_l^* = \frac{dY^*_{\varepsilon,l}}{dX}\left[\frac{dX^*_{\varepsilon,l}}{d\gamma}\Delta\gamma + \frac{dX^*_{\varepsilon,l}}{d\beta}\Delta\beta\right] + \frac{dY^*_{\varepsilon,l}}{d\gamma}\Delta\gamma$$

$$= \left[\eta_l f_{X^E}\left(X^{E*}_{\varepsilon,l}\right)\right]\left(-D^{-1}\eta_l\gamma_\varepsilon\left[p_o f_{X^E}\left(X^{E*}_{i,l}\right)[1-\phi]\Delta\gamma - p_e\Delta\beta\right] + X^*_{\varepsilon,l}\Delta\gamma\right)$$

$$\geq 0 \; iff \; \phi \leq 1 - \left(\frac{p_e\Delta\beta}{p_o\Delta\gamma}\right)\frac{1}{f_{X^E}\left(X^{E*}_{i,l}\right)} + \left(\frac{DX^*_{\varepsilon,l}}{\eta_l\gamma_\varepsilon}\right)\left[p_o f_{X^E}\left(X^{E*}_{i,l}\right)\right]^{-1}$$

Derivation of eqn (31)

Note that eqn (29) can also be written as:

$$\Delta B_l = p_o \Delta Y_l^* - \Delta X_l^* p_{\varepsilon,l}^{wA} - K_1 - p_e \eta_l \Delta X_l^{B*}$$

$$\Delta X_l^{B*} = \frac{dX_{\varepsilon,l}^{B*}}{d\gamma}(\gamma_1 - \gamma_0) + \frac{dX_{\varepsilon,l}^{B*}}{d\beta}(\beta_1 - \beta_0)$$

$$= \beta_\varepsilon \frac{dX_{\varepsilon,l}^*}{d\gamma}(\gamma_1 - \gamma_0) + \left[X_{\varepsilon,l}^* + \beta_\varepsilon \frac{dX_{\varepsilon,l}^*}{d\beta}\right](\beta_1 - \beta_0) = \beta_\varepsilon \Delta X_l^* + X_{\varepsilon,l}^*(\beta_1 - \beta_0)$$

$$\Delta B_i = p_o \Delta Y_i^* - \Delta X_i^* \left[p_{i,\varepsilon}^{wA} + p_e \beta_i \eta_\varepsilon\right] - Z_1 - p_e \beta_i X_{i,\varepsilon}^*(\eta_1 - \eta_0)$$

$$= \Delta X_i^* \left\{p_o f_{X^E}\left(X_{i,\varepsilon}^{E*}\right)\eta_\varepsilon \gamma_i - p_e \beta_i \eta_\varepsilon - p_{i,\varepsilon}^{wA}\right\} + X_{i,\varepsilon}^*(\eta_1 - \eta_0)$$

$$[p_o \gamma_i - p_e \beta_i] - Z_1$$

$$= X_{i,\varepsilon}^*(\eta_1 - \eta_0)[p_o \gamma_i - p_e \beta_i] - Z_1$$

because the coefficient on ΔX_i^* in the second line above is evaluated at an optimum and thus equal to 0 by eqn (3).

Derivation of eqn (32)

Note that eqn (30) can also be written as:

$$\Delta B_i = p_o \Delta Y_i^* - \Delta X_i^* p_{i,\varepsilon}^{wA} - Z_1 - p_e \beta_i \Delta X_i^{A*}$$

$$\Delta X_i^{A*} = \frac{dX_{i,\varepsilon}^{A*}}{d\eta}(\eta_1 - \eta_0) = \left[\eta_\varepsilon \frac{dX_{i,\varepsilon}^*}{d\eta} + X_{i,\varepsilon}^*\right](\eta_1 - \eta_0) = \eta_\varepsilon \Delta X_i^* + X_{i,\varepsilon}^*(\eta_1 - \eta_0)$$

$$\Delta B_i = p_o \Delta Y_i^* - \Delta X_i^* \left[p_{i,\varepsilon}^{wA} + p_e \beta_i \eta_\varepsilon\right] - Z_1 - p_e \beta_i X_{i,\varepsilon}^*(\eta_1 - \eta_0)$$

$$= \Delta X_i^* \left\{p_o f_{X^E}\left(X_{i,\varepsilon}^{E*}\right)\eta_\varepsilon \gamma_i - p_e \beta_i \eta_\varepsilon - p_{i,\varepsilon}^{wA}\right\} + X_{i,\varepsilon}^*(\eta_1 - \eta_0)$$

$$[p_o \gamma_i - p_e \beta_i] - Z_1$$

$$= X_{i,\varepsilon}^*(\eta_1 - \eta_0)[p_o \gamma_i - p_e \beta_i] - Z_1$$

because the coefficient on ΔX_i^* in the second line above is evaluated at an optimum and thus equal to 0 by eqn (3).

Chapter 12

Toward sustainable irrigation: Would subsidizing improve technologies result in water conservation in Alberta, Canada?

Lorraine Nicol[1], Henning Bjornlund[1,2] & K.K. Klein[1]
[1]*Department of Economics, University of Lethbridge, Lethbridge, Alberta, Canada*
[2]*Also University of South Australia, Adelaide, Australia*

Abstract

In semiarid southern Alberta where irrigation accounts for the majority of water consumed, improved efficiency in this sector has been advocated as a means to reduce water use and increase availability to new users, including in-stream users. This chapter evaluates the importance of improved irrigation technologies and management practices in improving water use efficiency and assessing whether subsidizing the improved technology is likely to reduce water use and promote more sustainable irrigation. The results find that water savings do not necessarily translate into water conservation as the saved water can be used to intensify irrigation, resulting in reduced return river flows. However, subsidy programs linked to reduced irrigated areas, and thereby water conservation, might work in southern Alberta since the main driver of adoption is cost saving, quality improvement or increased yield while very little importance is placed on increasing the irrigated area.

Keywords: Irrigation; Irrigation efficiency; Irrigation technology; Irrigation water savings; Subsidizing irrigation technology

1 Introduction

Alberta, and especially the semiarid area of southern Alberta, recently has started to notice the impacts of increasing demands for water, which exerts pressure on an already scarce resource. The seriousness of the situation became clear when the province disallowed any increased extraction of water for consumptive use within nearly every river system in southern Alberta while it searched for solutions. A long-term strategy document for water management was released in 2003 (Alberta Environment (AENV), 2003). Entitled "Water for Life", the document notes that current and future demand for water to ensure economic growth, support a growing population, and secure healthy rivers and lakes eventually may exceed available supply. The document proposes a strategy of water conservation that could be achieved through a 30% increase in water use efficiency and productivity. The conserved water should then move to satisfy the increased demand from other sectors of the economy and the environment by voluntary reallocations. The strategy document emphasizes the sustainability of aquatic ecosystems to ensure sustainable economic development, but the means by which these objectives are to be achieved are not made explicit.

Irrigators' contributions toward the strategy's objectives will be imperative as irrigation in Alberta accounts for 71% of consumptive use of surface water in the province (Alberta Environment (AENV), 2002). Improved efficiency of water use in agriculture has been advocated as a means to reduce water consumption among existing users and to increase the supply available for new users, including in-stream uses (Alberta Environment (AENV), 2003). However, as noted in Chapter 1, care must be taken to design programs so that efficiency gains do not result in a net increase in water use. The problem with improving water use efficiency in irrigation was highlighted by Whittlesey (2003) and English *et al.* (2002) when they found that the main purpose of adopting improved technologies for water application is to increase yields, and not to save water. Often the result of the introduction of more efficient technology is an increase rather than a decrease in water consumption, reducing return flow for downstream users and ultimately representing a tradeoff between benefits derived from upstream agricultural uses and benefits from downstream extractive and in-stream uses (Whittlesey, 2003).

Return flows vary significantly across irrigation districts in southern Alberta. A 5-year monitoring program from 1996 to 2000 found that return flows range from as low as 7% to as high as 56% of gross diversion (AIPA, 2002); hence, reductions in these return flows could have significant consequences for river flows. Whittlesey (2003) advised against practices such as subsidizing adoption of more efficient irrigation technologies. Scheierling *et al.* (2006) found that such subsidies resulted in water conservation only if they are subject to commitments to reduce the irrigated area, reduce the number of irrigation events or change to less water-intensive crops. Their results suggest that in areas where the main drivers of adoption are cost reduction in energy and labor or improved quality of output, irrigators will be most likely to accept such conditions imposed on subsidization programs.

Water use efficiency in Alberta, measured as the fraction of water delivered to the farm that actually reaches the root zone of crops, has improved over time through the gradual movement away from surface irrigation (with an average of 30% water use efficiency) to wheel move irrigation and ultimately to low-pressure center pivots (with an average of 80% water use efficiency). Under existing irrigation techniques, overall water use efficiency was estimated in 2007 at 72.5% across all irrigation districts, compared to only 34% in 1965 (Alberta Irrigation Projects Association (AIPA), 2009).

This chapter evaluates the importance of improved irrigation technologies and management practices in improving water use efficiency in irrigation in southern Alberta and assesses whether subsidizing the improved technologies is likely to reduce water use and promote more sustainable irrigation. In Alberta, irrigators have the consumptive right to all the water they are licensed to divert (Water Act, 1999). Therefore, it is at the irrigators' discretion how any water saved through adopting improved technology is used. The irrigator might (1) divert less and thereby increase river flows, (2) increase the irrigated area or (3) sell the water (Nicol & Klein, 2006; Nicol *et al.*, 2008).

2 Policy and legislative framework

Balancing economic and ecosystem water requirements is a major challenge in southern Alberta. Rapid economic and population growth have taken their toll on the river systems. Assessment of the condition of the river reaches within the South Saskatchewan River Basin (SSRB), where most population growth and economic and irrigation activity occur, indicate that 22 of 33 main stem river reaches are moderately affected, five reaches are heavily affected, and three are degraded. The status of the three other reaches overlapped categories in the grading system (AENV, 2005).

Sustainable irrigation is integral to sustainable water management in southern Alberta because of the great amount of water used for that activity. Irrigation's integration in the province's overall water management is embedded in the Water for Life strategy's objective of "reliable quality water supplies for a sustainable economy" (AENV, 2003). Although vague on detail, the Water for Life strategy indicates that economic instruments will be used as means to provide incentives toward the achievement of its increased efficiency and productivity goals. Aside from subsidies and water markets already in place and discussed later in this chapter, future economic instruments could be broadened to include pricing, levies, or taxes. A major requirement is for all sectors, including irrigation, to demonstrate best management practices and improved water use efficiency and productivity by 2015 (Alberta Water Council (AWC), 2008). The irrigation industry is currently in the process of devising plans to do so.

Fundamental changes to Alberta's water legislation took place with the passage of the Water Act in 1999, followed by the Irrigation Districts Act in 2000. The Water Act is the overarching legislative tool in managing water to

> ". . . . support and promote the conservation and management of water, including the wise allocation and use of water, while recognizing (a) the need to manage and conserve water resources to sustain our environment and to ensure a healthy environment and high quality of life in the present and the future ," (p. 14)

Unlike the top–down, government-directed approach to water management embodied in earlier legislation, the management paradigm is now integrated, shared and cooperative (Water Act, 1999). The Act also protects the seniority of existing water license holders who are in good standing, prohibits the export of Alberta's water to the United States, and prohibits interbasin transfers of water between Alberta's major river basins.

Driven by the spirit of the Water Act, local stakeholders' involvement and the devolution of water management to watershed levels are occurring through a network of partnerships, consisting of three facets (AWC, 2008):

1. Alberta Water Council – responsible for the development of strategic policy at the provincial level.
2. Watershed Planning Advisory Council (WPAC) – responsible for planning at the watershed or basin level. Each WPAC is a stand-alone, incorporated society with a mandate for effective water management in its watershed (AWC, 2005). Their principal mandates are to develop state of the watershed reports and watershed management plans (AWC, 2008).
3. Watershed Stewardship Groups – perform a combination of grassroots work, public education, and engagement activities.

The overarching purpose of watershed planning is to "resolve water management issues such as the availability of water for future allocations and river flows needed for protection of the aquatic environment" (AENV, 2002, p. i). By late 2008, the Alberta Water Council, 8 WPACs and over 140 Watershed Stewardship Groups had been established (AWC, 2008). It is common for the WPACs to have a large number of members who are grappling with balancing the triple bottom line of ecological, social, and economic benefits of water. Lack of resources, confusion over roles and responsibilities, volunteer burnout, and variance in expertise and knowledge among members have been cited as contributing to slow progress (AWC, 2008; Poirier, 2008; Cryderman, 2008). One report noted that setting instream flow levels has, in general, proved to be a controversial aspect of implementing local basin management plans (Conference Board of Canada, 2008). While developing a watershed management plan that includes water conservation objectives is a critical and primary responsibility of the WPACs, they do not have any powers of implementation.

Alberta is the only province in Canada that has the ability to transfer water independent of land. The provision is seen as a "mechanism to provide economic efficiency and flexibility in basins where water resources are fully allocated" (Water and Duncan Gordon Foundation (WDGF), 2007, p.16). Under the Water Act, both transfers of water allocations on a temporary basis and water rights on a permanent basis are permitted. Under the Irrigation Districts Act, the owner of irrigated land can transfer water licenses to other irrigators within the same irrigation district (with the approval of the District's Board of Directors). Transfers of all or a portion of a license outside the district is possible, but only if a plebiscite is held and a majority of irrigators agree.

Although government views markets for water as being fundamental to reallocation of water to meet future demands, markets for water rights have been slow to develop in Alberta. Anecdotal evidence suggests that transfers between users on a temporary basis are nonexistent in years when rainfall is near normal or higher. This mechanism did, however, prove useful during the 2001 drought year when water allocations were relatively easy for farmers to access within the St. Mary River Irrigation District (SMRID) (Nicol, 2005; Nicol & Klein, 2006).

Fewer than a dozen transactions involving the permanent trade of water rights have been completed in southern Alberta (Nicol *et al.*, 2008). Some cases have proved to be so controversial that the provincial government has been forced to put in restrictions and safeguards. Several administrative factors have made it difficult for buyers and sellers of water rights to negotiate and complete transactions. These include cumbersome administrative procedures, lack of mechanisms for buyers and sellers to locate each other, and no publicly available information on market prices. A high-profile case to obtain a permanent license to 2,500 dam^3 of water for a commercial complex in 2007 caused significant controversy and public outcry, underscoring the fact that not everyone agrees that water management should be left to market forces. The water finally was secured from the Western Irrigation District in return for $15 million to convert a leaky canal into a 50-km pipeline (D'Aliesio, 2007). The plebiscite held in the irrigation district passed narrowly with 57% of the 328 voters approving the transaction, a difference of only 46 votes. Public opposition and an (unsuccessful) appeal to the Environmental Appeal Board underscored the high degree of discontent.

Opposition to water markets also occurred when the government was forced to halt the practice of amending irrigation district licenses to enable the districts to sell water for uses other than agriculture. A government review of this practice likely will result in restrictions, including capping the amount of district irrigation licenses that can be sold for purposes other than irrigation and adding a requirement for an environmental review of each case. Government, therefore, has been forced to place restrictions on water market activity to satisfy conflicting groups, mainly between environmental, agriculture, and urban interests.

3 Irrigation in Alberta and Canada

Nearly two-thirds of irrigation activity in Canada is located in Alberta (Statistics Canada). Irrigation is by far the major water consumer in Alberta, and 82% of the irrigated area is located within the SSRB, in the semiarid southern region.

In Alberta, water is managed under a licensing system. Licenses are legallytied to specific parcels of land and historically have remained with the property when the land is sold. The date the license was issued establishes its seniority, and the first-in-time, first-in-right principle ensures that during times of water shortages, license holders obtain access to their water in accordance with their seniority.

For irrigation purposes, licenses are held by two distinct types of irrigators: private and district irrigators (Figure 1). Private irrigators are issued individual licenses just like other private or public landowners who need water for purposes, such as golf courses or parks. Licensing of private irrigators is regulated by Alberta Environment. Almost 2900 private irrigators exist in Alberta, accounting for 18%

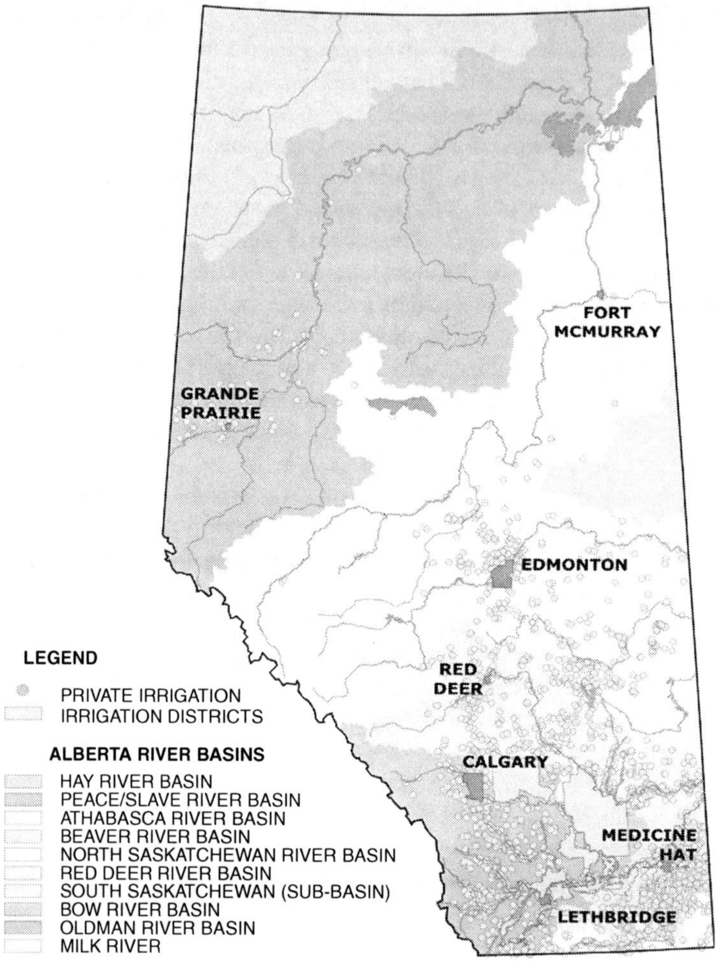

Figure 1: Private and district irrigation. Adapted from Alberta Irrigation Projects Association (AIPA). 2009. Irrigation Sector – Conservation, Efficiency, and Productivity Plan (Draft). AECOM Canada Ltd.
Source: Alberta Agriculture and Rural Development.

of the irrigated area (Alberta Agriculture, Food and Rural Development (AAFRD), 2007). Water for private irrigation is extracted mainly from 10 rivers within the SSRB. Private irrigators are responsible for the installation and maintenance of the infrastructure needed to pump the water from the river and convey it to the field, and the irrigation equipment used to apply the water. Private irrigators do not pay for the water. When a license is approved, a one-time levy must be paid based on the volume of water involved. These irrigators are governed by the Water Act (1999).

The provinces' 13 irrigation districts, which account for 82% of the irrigated area, operate very differently from private irrigators. The irrigation districts hold water licenses and the farmers who belong to the districts have their irrigable area on the district's assessment role. These irrigators constitute the district's ratepayers. Irrigators pay a flat fee per hectare for administration costs and 25% of the cost of rehabilitating the district infrastructure. The annual assessments vary from \$18.50 ha^{-1} in the Eastern Irrigation District to almost \$45 ha^{-1} in the SMRID (AAFRD, 2007). This variation reflects differences in costs incurred by the irrigation districts to provide the water and whether they have access to other sources of income.

Like private irrigators, district irrigators do not pay for the water themselves and neither type of irrigator pays for the cost of head works and the supply infrastructure used for delivering the water to the point where the district or irrigator extracts the water from the river. Since 1969, the Alberta government has subsidized irrigation by paying for the majority of the costs of rehabilitating the district infrastructure (AIPA, 2002). Some irrigation districts supply water to municipalities, golf courses, feedlots, and oil, gas, and other industries, resulting in a complex fee structure among districts.

4 Survey design and methods

In 2005, the managers and board members of the 13 irrigation districts were surveyed to determine the irrigation industry's perception of the Water for Life strategy and its main objectives and policy instruments (Bjornlund *et al.*, 2007). The study found that there was very little support for the use of economic instruments and the average expectation of further efficiency gains was far below the 30% target outlined in the Water for Life strategy document. It was found that further adoption of improved technologies or management practices likely would vary across the 13 districts depending on physical production conditions and other factors. The survey also revealed that the willingness by the irrigation sector to transfer water to other sectors including the environment is very low. Many in the irrigation sector believe that saved water should go to expand irrigated production or to secure existing production during periods of drought.

The survey was followed by two additional surveys that obtained the views and practices of private and district irrigators within two irrigation districts – Taber and Raymond. These surveys delved more deeply into issues relating to water use efficiency measures and reaction to the use of economic instruments. The survey asked irrigators: (1) what they have done in the past and what they intend to do in the future to improve water use efficiency on their farms; (2) what were the drivers and impediments to make such improvements; (3) what has influenced their decision making about improved technologies and management practices; and (4) what are their likely responses to economic instruments including subsidies, greater opportunities to obtain specialty crop contracts, and trading water allocations and water rights. Many of the questions sought information on irrigator's historical, recent past, and future intended technological and management practices in three distinct time periods: historical (before 2001), recent past (2001–06), and future (2007–12). More detailed discussions of the two surveys are provided in Bjornlund *et al.* (2009) for district irrigators and Nicol *et al.* (2008) for private irrigators.

5 Survey results

5.1 Property and farmer characteristics

The studies revealed significant differences in production and personal characteristics between private and district irrigators. A factor of particular importance, one likely to affect virtually all decision making, is the prominence of dry land farming among private irrigators (Table 1). A significantly higher percentage of private irrigators have large amounts of land under dry land farming compared to district irrigators – 39% of private compared to 16% of district irrigators have more than 260 ha of dry land (Pearson's chi-square $p<0.01$). Alternatively, only 13% of private irrigators have irrigated areas greater than 260 ha compared to 24% of district irrigators (Pearson's chi-square $p<0.05$). This suggests that private irrigators are less reliant on irrigation to generate farm revenue and the cost of irrigation and the efficiency of irrigation therefore could be of less importance to them.

There was also a noticeable difference in crops produced between the two groups. A significantly higher proportion of private irrigators grow forage (Pearson's chi-square $p<0.01$). Ninety-three percent of private irrigators indicated that they use their relatively small irrigated area to grow forages in support of their primary production – mainly cow-calf and feedlot operations. Fifty-five percent of private irrigators indicate that all their irrigated area is dedicated to forage production. So prominent is forage production that 64% of private irrigators indicate they have no irrigated area dedicated to cereal production and 87% have no area dedicated to specialty crop production. While 31% of Taber and Raymond Irrigation District farmers dedicate all their area to forage, for most, forage is the final product and not an input to a primary value-added production. A significantly higher proportion of district irrigators had more land in specialty crops and cereals (for both, Pearson's chi-square $p<0.01$). Approximately 40% indicated they have some irrigated area under both cereals and specialty crop production,

Table 1: Dry land and irrigation farming.

	Dryland[1]		Irrigation[2]	
Size (hectares)	Private	District	Private	District
<65	38	60	55	42
65 and <130	9	16	20	16
130 and <260	14	9	13	18
260 and >	39[1]	16	13	24

[1]Significantly different at the 0.01 level;
[2]Significantly different at the 0.05 level.
Figures reported are percentage of all respondents.

reflecting the need for crop rotations on their farms. A significantly higher proportion (43%) of private irrigators reported that the weather is the most important factor determining when to irrigate, while a significantly higher proportion (57%) of district irrigators reported that the crop growth stage is the most important factor. This indicates that district irrigators are more intensive irrigators, with irrigated crops accounting for most of their revenue. They are therefore likely to be more concerned about the cost associated with irrigation and the efficiency of water use. Since their crops that are irrigated tend to be final products for sale and many are growing specialty crops for processing with strict quality parameters to fulfill contracts and achieve high prices, they also were more likely to be influenced by the quality gains that are associated with the adoption of some new irrigation technology.

Comparing additional production and personal characteristics between the two groups indicates that, compared to district irrigators, private irrigators seem to:

- have significantly more land under less advanced surface ($p<0.01$) and wheel move irrigation systems ($p<0.05$); 21% have all their irrigated land under surface irrigation compared to 8% of district irrigators, 28% have all their irrigated land under wheel move irrigation compared to 21% of district irrigators, 5% have all their irrigated land under high-pressure pivot compared to 8% of district irrigators ($p<0.01$),
- have less off-farm work (52% compared to 61%, $p<0.1$) and depend less on off-farm income (21% derive over 75% of household income from off-farm work compared to 40%, $p<0.1$),
- have less formal education (29% have college education compared to 32%, 6% have a university graduate degree compared to 14%, $p<0.1$),
- are significantly older (67% are 55 years or older compared to 45% of district irrigators, $p<0.01$),
- have a significantly longer family background in farming (91% compared to 84% of district irrigators, $p<0.05$)
- have significantly greater expectation of family continuity of the farm (33% compared to 48% of district irrigators do not expect family continuity, $p<0.01$),

These differences, combined with the differences in farm size, cropping, and reliance on irrigation, suggest that private irrigators are financially more stable enterprises, less exposed to risk due to investment in irrigation infrastructure, and less sensitive to price fluctuations and quality demanded of their irrigated products. They are more likely to have properties passed down in the family and have a greater expectation of continuing this tradition, and also are less dependent on off-farm work. This all suggests that they might be financially more viable. The low level of dependence of irrigation, and almost sole use of irrigated production in their own value-added enterprises, suggest that private irrigators might not be too concerned about improving irrigation technology and management. On the other hand, they seem to be more financially capable of financing technology changes, and may do so more readily with subsidization.

5.2 Adoption of improved technologies

Relative to district irrigators, private irrigators have invested less in new irrigation technologies in the past and have less intention to invest in the future. As Table 2 shows, across most initiatives, a significantly lower percentage of private irrigators have undertaken these measures compared to district irrigators. For the time periods of before 2001 and 2001–06, the percentage of district irrigators implementing these measures was significantly higher, in many cases more than double the percentage of private irrigators. The rate of adoption for three measures slowed considerably for both groups from the first to the second period – converting from surface to wheel move, wheel move to pivot, and surface to pivot. This probably reflects that most farmers who intend to, or can afford to, make these changes, had already done so by 2001. It also reflects that by 2006 these technologies had been around for a long time. The percentages for converting from high to low pressure and purchasing a computer panel increased for both groups. This may be because these are two recent irrigation technologies used in Alberta, with adoption further prompted by increasing energy costs, increasing labor, problems associated with getting labor, and increasing demand for quality products by processors. In the future, producers in both groups indicated that the rate of adoption is likely to slow considerably, with the highest percentage indicating that they plan to change to low-pressure pivots and purchase a computer panel. These percentages are low but district irrigators seem likely to adopt these measures at a significantly faster pace than private irrigators. This is a reflection that district irrigators are far more dependent on irrigation as many grow specialty crops. They are therefore sensitive to increased costs associated with irrigation and the need for precision with which irrigation water is delivered and its impact on product quality.

Table 2: Implementing irrigation technologies.

Type	Before 2001		2001–06		2007–12	
	Private	District	Private	District	Private	District
Surface to wheel move	20[1]	37	1[1]	6	2	3
Wheel move to pivot	13[1]	33	7[2]	20	2[1]	10
Surface to pivot	7	10	1	3	1	2
High to low pressure	5[1]	16	10[1]	23	3[1]	12
Purchase computer panel	5	7	6[1]	16	1[1]	12

[1]Significantly different at the 0.01 level;
[2]Significantly different at the 0.05 level.
Figures are reported in percentage of respondents.

5.3 Reasons for and against adoption

Irrigators were asked to rate how important they found a number of reasons for improving water management and then identify which one they perceived to be the most important (Table 3). Far more district irrigators found all the reasons to be important in their decisions, both relating to reducing cost and increasing volume and quality of crops. When asked to identify which was the most important, both groups rated the reasons in the same order of importance. By far the most important for the district irrigators was to improve crop yield and quality. For private irrigators, almost the same proportion found three of the reasons to be the most important – two related to reducing inputs (energy and labor) and the third to improving crop yield and quality. These differences clearly illustrate the higher level of dependence on irrigation for intensive specialty crop production among the district irrigators. Their return is dependent on the volume and quality of their irrigated production. On the other hand, the private irrigators do not rely to the same extent on sales of their irrigated crops but use them as inputs into the production of their final products. Consequently, a significantly higher proportion is most interested in controlling cost and labor inputs required. This difference also is caused by the fact that pumping costs have been estimated to be twice as high for private irrigators as for district irrigators, due to associated higher pump lifts to divert water to their fields (AIPA, 2009). It also should be noted that a much higher proportion of district irrigators rate all the reasons much

Table 3: Reasons for improving irrigation water management.

	Private irrigators		District irrigators	
Reason	Important	Most important	Important	Most important
To improve crop yield and quality[1]	64	30	91	56
To reduce energy costs[1]	46	24	73	13
To reduce water use[1]	41	9	68	9
To reduce labor input[2]	55	27	69	16
To reduce fertilizer or pesticide losses[1]	14	1	63	1
To reduce soil erosion[1]	21	1	59	2
To irrigate more land with the same water[1]	26	3	73	1

[1]Significantly different at the 0.01 level;
[2]Significantly different at the 0.10 level.
Figures are reported in percentage or respondents.

higher than the private irrigators, confirming that irrigation is a far more important component of these farms.

When asked to rate how important they found a number of reasons influencing their decision not to improve their water management and to identify the most important reason, the differences between the two groups were not nearly as significant as were the reasons for adoption. District irrigators tend to believe that they already use all the water-saving practices that are practical. They also were concerned about finances. While poor commodity prices were rated as important by most, it was rated the most important by only 1%, suggesting that their poor financial situation is perceived to be more complex than simply poor commodity prices.

A significantly smaller number of private irrigators indicated that they already have all the water-saving devices that are practical. This suggests that private irrigators believe they have more room to improve, a not unexpected finding since a significantly lower proportion have or plan to invest in improved irrigation efficiency. For private irrigators, physical field conditions were the most important factors for not investing. While located closer to the natural watercourses, their property is more undulated and unsuitable for center pivots. It is also noteworthy that private irrigators believe improvements will reduce operational costs but not enough to cover installation costs, thereby causing them to aim more toward improving management.

Table 4: Reasons inhibiting improved irrigation water management.

	Private irrigators		District irrigators	
Reason	Important	Most important	Important	Most important
Already use all water-saving practices[2]	36	15	47	25
Financial situation does not permit the investment	30	14	39	21
Will reduce costs but not enough to cover installation costs	37	16	30	9
Physical field conditions limit system improvements	41	28	32	12
Poor commodity prices[1]	14	1	63	1

[1]Significantly different at the 0.01 level;
[2]Significantly different at the 0.10 level.
Figures reported are percentage of respondents.

6 Subsidizing new technologies

Irrigators were asked to choose among three options of financial assistance to assist them in investing in more efficient irrigation equipment. Of the three options – cash subsidy, subsidized interest rates, and accelerated depreciation – both private and district irrigators favored cash subsidies – 74% and 71%, respectively. In terms of the amount of assistance needed to do so, private irrigators appear to require greater financial incentives, reflecting that they also are more likely to believe that improvements will reduce operational costs but not enough to cover installation costs. Although the differences were not statistically different, almost 60% of private irrigators indicated they would require more than \$10,000 to improve existing equipment but only about one-third of district irrigators would require that much. To invest in a new, low-pressure pivot, close to 70% of private irrigators indicated they would require over \$30,000, representing less than a third of the cost of a new low-pressure pivot. A much lower percentage of district irrigators, about one-third, would require that level of subsidization (Table 5). About half of the district irrigators indicated they would require the mid-range level of subsidization. This seems relatively low and, combined with our findings that cost and labor reductions are major drivers of adoption, supports Scheierling *et al.*'s (2006) findings.

An ongoing subsidy program within the Eastern Irrigation District supports these conclusions. The district offers irrigators a subsidy of C\$17,020 to change from flood irrigation to a low-pressure center pivot system covering an area of 132 acres out of the 145 acres previously covered by flood irrigation (Wilson, Personal Communication). This amount is paid only if they reduce their irrigated area by the 13 acres not irrigated by the new center pivot system, or a reduction of 9%. This subsidy constitutes approximately 17% of the cost of the system. The district reports a high participation rate, resulting in an increase of 50% in area under low-pressure center pivot. If the irrigator decides not to have their irrigated area reduced, the subsidy will be only C\$4020, or 4% of the cost of the system. Half of the subsidies have resulted in a reduction in irrigated area, resulting in water savings of almost 10,000 dam^3. However, the water saved typically

Table 5: Level of subsidization based on 65lha.

Improve existing equipment			Invest in new low-pressure pivot		
Level	Private	District	Level	Private	District
<\$5,000	23	18	<\$10,000	12	20
\$5,000–10,000	18	47	\$10,000–30,000	16	49
>\$10,000	59	34	>\$30,000	68	31

Figures reported are percentage of all respondents.

has been sold by the district to irrigate additional land (Wilson, Personal Communication). Hence, the program has achieved two of the objectives of the Water for Life strategy, increasing water use efficiency and productivity, but not water conservation.

7 Conclusions

Alberta is in the early stages of experiencing water scarcity with predictions that water demand from all sectors including the environment will continue to increase. As a consequence, most rivers in the southern part of the province have been closed to the issuance of new licenses. With a new Water Act in 1999, Alberta started down the road toward sustainable water management. This was followed by a new Irrigation Districts Act in 2000 and the Water for Life strategy in 2003, which was renewed in 2008. A key component of the solution to achieving sustainable water management is increased water use efficiency and productivity, resulting in water savings and conservation that can be reallocated to new users, including the environment. As irrigation accounts for around 71% of all water consumed in the SSRB (AENV, 2002), agriculture necessarily will have to play a pivotal role in this process, with sustainable irrigation being a primary objective.

This chapter demonstrates that most of the improvements in irrigation technologies that can be expected to be implemented by the current generation of farmers, and under current commodity prices, already have taken place. It also confirms the findings of other researchers that water savings in agriculture do not necessarily result in water conservation as the saved water can be used to intensify or expand irrigation, resulting in a reduction in river flows. Also, as indicated by other researchers, our findings suggest that subsidy programs linked to reduction in irrigated areas, and thereby water conservation, might work in Alberta since the main drivers of adoption are cost saving, improved crop quality, and increased yield, while less importance is placed on increasing the irrigated area. This is more so among private irrigators than among district irrigators. However, a current subsidy program linked to reductions in irrigated area suggests that a large number of district irrigators are willing to see their irrigated area reduced as part of a subsidy package. The findings, however, also clearly indicate that to ensure water conservation as source of supply for new users and the environment, regulatory instruments are likely to be needed.

References

Alberta Agriculture, Food and Rural Development (AAFRD). 2007. Alberta Irrigation Districts—Alberta Irrigation Information for the Year 2006, Lethbridge, Alberta: Author, 2007.

Alberta Environment (AENV). 2002. *Water for Life – Facts and Information on Water in Alberta.* publication #I/930. Edmonton, Alberta: Author, 2002.

Alberta Environment (AENV). 2003. *Water for Life – Alberta's Strategy for Sustainability*. Edmonton, Alberta: Author, 2003.

Alberta Environment (AENV). 2005. Background Information for Public Consultation on the South Saskatchewan River Basin Draft Water Management Plan, 2005. Edmonton, Alberta: Author, 2005.

Alberta Irrigation Projects Association (AIPA). 2002. Irrigation in the 21st Century, Volume 1: Summary Report, 2002. Lethbridge, Alberta: Author, 2002.

Alberta Irrigation Projects Association (AIPA). 2009. Irrigation Sector – Conservation, Efficiency, and Productivity Plan (Draft). AECOM Canada Ltd.

Alberta Water Council (AWC). 2008. Water for Life – A Renewal.

Alberta Water Council (AWC). 2008. What We Heard: Summary Findings of the Shared Governance-Watershed Management Planning Workshops.

Alberta Water Council (AWC). 2005. Review of Implementation Progress of Water for Life, 2004/2005.

Bjornlund, H., Nicol, L. & Klein, K. 2007. Challenges in implementing economic instruments to manage irrigation water on farms in southern Alberta. *Journal of Agricultural Water Management,* 92: 131–141.

Bjornlund, H., Nicol, L. & Klein, K. 2009. The adoption of improved irrigation technology and management practices—a study of two irrigation districts in Alberta, Canada. *Agriculture Water Management,* 96: 121–131.

Conference Board of Canada. 2008. Going with the Flow? Evolving Water Allocations and the Potential and Limits of Water Rights Transfers in Canada. Draft report, prepared by Brandes, O., Nowlan, L., Paris, K. & Wilts, R.

Cryderman, K. 2008. Faster Alberta Action on Water Sought. Calgary Herald, p. A9, November 20, 2008.

D'Aliesio, R. 2007. Tapped Out. The Clash Over Alberta's Water. The Calgary Herald, pp. A19–22, December 1, 2007.

English, M., Solomon, K. & Hoffman, G. 2002. A paradigm shift in irrigation management. *Journal of Irrigation and Drainage Engineering,* 128(5): 267–277.

Nicol, L. 2005. Water Markets in Southern Alberta. M.A. Thesis, University of Lethbridge, Lethbridge.

Nicol, L. & Klein, K. 2006. Water market characteristics: results from a survey of Southern Alberta irrigators. *Canadian Water Resources Journal,* 31(2): 91–104.

Nicol, L. Bjornlund, H. & Klein, K. 2008. Improved Technologies and Management Practices in Private Irrigation—Implications for Water Savings in Southern Alberta. Paper presented at the Water Down Under conference, Adelaide, Australia, April, 2008.

Nicol, L., Klein, K. & Bjornlund, H. 2008. Permanent transfers of water rights: a study of the southern Alberta market. *Prairie Forum,* 33(2): 341–356.

Poirier, B. 2008. Assessing Institutions for Aquatic Ecosystem Protection: A Case Study of the Oldman River Basin, Alberta. PhD Thesis, University of Guelph, ON, Canada.

Scheierling, S., Loomis, J.B. & Young, R.A. 2006. Public subsidies for water-conserving irrigation investments: hydrologic, agronomic, and economic assessment. *Water Resources Research,* 42: 1–11.

Statistics Canada, 2006. Snapshot of Canadian Agriculture: http://www.statcan.gc.ca/ca-ra2006/articles/snapshot-portrait-eng.htm.

Water Act. 1999. http://www.environment.alberta.ca/3.html, CFM.

Walter and Duncan Gordon Foundation (WDGF). 2007. Water Allocation and Water Scarcity in Canada: Initiating a Policy Dialogue for the 21st Century. Prepared by de Loe, R., Varghese, J., Ferreyra, C. & Kreutzwiser, R.

Whittlesey, N. 2003. Improving irrigation efficiency though technological adoption: When will it conserve water? In: Alsharhan, A.S. & Wood, W.W (Eds.), *Water Resources Perspectives: Evaluation, Management and Policy.* Elsevier Science, Amsterdam, pp. 53–62.

Wilson, E. Personal communications 5 and 20 February, 2009, General Manager Eastern Irrigation District.

Chapter 13

Application of economic instruments, tradable licences and good governance for sustainable irrigation water conservation in South Africa

W.L. Nieuwoudt[1] & G.R. Backeberg[2]
[1]*University of KwaZulu-Natal, Pietermaritzburg, South Africa*
[2]*Water Research Commission, Pretoria, South Africa*

Abstract

According to the National Water Policy in South Africa, the objectives are to achieve equitable access to water, and to promote efficient and sustainable use for optimum social and economic benefit. The National Water Act provides the legal framework, and the National Water Resource Strategy explains the ways in which water resources will be managed. Several interrelated actions are taken to implement water conservation and demand management. These include levying of charges for water resource management, water resource development, water use and waste discharge; regulations to encourage and enforce water measurement; adoption of administrative and market processes for the allocation of water use entitlements and decentralization of authority for water management to catchment management agencies (CMAs). The results of completed and ongoing research projects in South Africa demonstrate the application of technologies for direct and indirect water measurement in rivers, canals and pipelines. Decision support systems are available to determine the cost of irrigation and assess the risks of agricultural water management on farms. Empirical analyses have been done for temporary or permanent transfers and lease or trade of water use entitlements on irrigation schemes with stable and variable water supply. The establishment and functioning of CMAs for local water management by different water use sectors including irrigation have been evaluated. On the basis of these studies, it is concluded that volumetric charging and cost recovery of water supply services through a two-part charging system are practically feasible. Complete description of water use entitlements and provision of information must improve to increase security and reduce the risks associated with market trades. Training and capacity building are essential to support participation by all water users and maintain standards of accountability and effective governance for irrigation management.

Keywords: Water governance; Water marketing; Water user charges

1 Introduction

Several studies have been undertaken to assist with the implementation of the new National Water Act (NWA) of 1998 (NWA, 1998) and the National Water Resource Strategy (NWRS) of 2004 (NWRS, 2004) by the Department of Water Affairs and Forestry (DWAF). The NWA brought about a new framework for water resource management in South Africa. Water resources should accordingly be allocated in a way that will ensure its 'best possible use'. The 'best possible use' entails more than the productive use of water since, in addition, social, economic and environmental factors must be included to achieve the objectives of equity, efficiency and sustainability of water use. These issues will be analysed and discussed in this chapter. The chapter is mainly based on completed and ongoing research projects managed and funded by the South African Water Research Commission.

2 Some theoretical and practical considerations

Water markets are based on a system of water law that displays three attributes: security, stability and flexibility in protecting transferability of property rights (Hobbs *et al.*, 1994). Security is the ability to identify and gain protection for the right of use. Stability assumes that the right of use will continue to be recognized. Flexibility allows the right of use to be transferred to another use. The flexibility of being able to transfer a water right adds value to it because the market value of the right reflects not only the value of current use but also that of future opportunities. Security and stability in water markets are important, as it will affect investment. A recent study indicated that irrigation farmers in the Lower Orange River, who view entitlements as less secure, expect to invest less. The issue of flexibility was discussed at a recent (2007) meeting with farmers in the Berg River. It appears that farmers near the river were reluctant to transfer water entitlements away from the river while farmers further away were keen to purchase more use rights. As demand for water is expected to increase in the future, demand for transfers away from the river may increase. By stating 'best use', the NWA has moved away from riparian use and these transfers are in accordance with this principle.

An economic explanation of 'efficient use' will be given as it was attempted to study this empirically in projects reported in this chapter. The economic interpretation of efficient use of water is that the return per cubic metre of water must be maximized. The full economic cost of water consists of financial, opportunity and external costs (Rogers as quoted by Perret and Geyser 2007). In a water market, the rent return to water is maximized as water moves to a better use. In the rent return to water, risk is reflected as a cost (opportunity cost). The economic meaning of efficiency of water use and the rent return to water are thus synonymous concepts. Supply risk is often high where not sufficient storage dams exist. This is important in South Africa as in the irrigation areas of the Crocodile River (East), water use rights moved from a high-risk environment to an environment where lower risk crops such as sugar cane can be produced.

It is also often stated that water markets are not effective because of few trades. This statement may miss probably the biggest contribution of a water market: It discovers the opportunity cost of water and all water users, including the non-buyers/sellers, are faced with this opportunity cost and are provided incentives to conserve water resources. However, participation in the water market is only possible by water users who have existing rights and have access to funds. Because of the history of South Africa, unequal allocation of use rights was caused by racially discriminatory legislation. This unequal distribution cannot be corrected in isolation by the market process. Political and legal negotiations are necessary to achieve equity through compulsory licensing as part of water allocation reform (Backeberg, 2007).

3 Economic instruments for water conservation

3.1 Charges for water resource development and use

A user charge is a non-market economic instrument, similar in nature to regulation or administrative control of a resource, and is used to encourage the conservation of water resources. The charge is levied by government or a catchment management agency (CMA) or a water user association (WUA) to recover the costs for providing a service to supply the allocated water. There is an exchange relationship between, for example, government as 'supplier' of the service and the user as the 'buyer'. In the case of a quasi-collective service such as the supply of irrigation water, the farming public pays a 'price' for such a service known as a user charge (Gildenhuys, 1997). Such instruments may ration or change resource use in the intended direction but is classified as ineffective if the market for allocation of water resources is not functioning. The reason is that incentives to increase efficiency that lead to water savings will only materialize if the water use entitlement is tradable.

The NWA provides for three types of water user charges to achieve water conservation and demand management (NWRS, 2004): (a) water resource management charges to fund the controlling, monitoring and protection of water resources in a catchment; (b) water resource development charges which recover the cost of planning, designing, constructing, operating and maintaining water supply schemes and (c) charges for achieving equitable and efficient water allocation which relates to the value of water. The third kind of charge has not been applied administratively (Genesis Analytics, 2005). The purpose of the first two financial charges is cost recovery while the objective of the third economic charge is purportedly to provide incentives to shift water from lower to higher value uses. The imposition of user charges such as under (a) and (b) is justified since the subsidization of irrigation schemes in the long run creates distortions by not relating actual costs to water supply. However, full cost recovery from farmers as direct beneficiaries is not justified since costs are also recovered indirectly through the taxation system (Gittinger, 1982). Charges to partially recover costs clearly make business and economic sense. For commercial farmers, the subsidy on operation and maintenance (O&M) costs has therefore been phased out. In contrast, for emerging and subsistence farmers, the O&M charges for water supply will be subsidized at a reducing scale over 5 years. Currently, user charges are levied on an area basis, but with measurement of water use, volumetric charges will be instituted. Further incentives to effectively manage both water and finances by WUAs can be achieved by means of a two-part charging system (Carruthers & Clark, 1981). This consists of a fixed and a variable component of water user charges for a specified storage and distribution capacity on irrigation schemes. Fixed charges include capital, interest, maintenance and administration costs, while variable charges are mainly to recover operating and repair costs.

The theoretical soundness of proposing a charge as envisaged under (c) to achieve equitable and efficient water allocation is questionable. Such a charge is in fact a tax which reduces the expected returns for productive use of water and is therefore rather a disincentive to transfer water. It is a fallacy to argue that economic charges will promote further conservation of water as resources are allocated based on opportunity cost and not financial cost. In fact, the user charges do not change the opportunity cost faced by the irrigator as the sum of the tax and the water rent (which is lower because of tax) will stay unchanged. Where the characteristics of a resource are conducive to the formation of a market, such as in the case of explicit, exclusive, enforceable and transferable water use entitlements, the most effective and efficient mechanism to promote resource conservation is to promote markets.

3.2 Costing of water use

In addition to user charges for water services, provision has to be made for on-farm cost of water abstraction, storage, distribution and application. These irrigation costs form a significant portion (15–27%) of the variable cost of commercial crop production (Backeberg, 2005). Before an investment decision is made, the capital and operating cost of irrigation equipment must therefore be evaluated. The program IrriCost has been developed to estimate the annual capital and operating cost of irrigation (Meiring *et al.*, 2002). This tool can be used to do cost comparisons of alternative designs, analyse annual cost of water use and compile budgets of irrigation costs. Apart from escalating costs, farmers are confronted with changing yields and prices. The model RiskMan was developed to provide information for risk management in irrigation farming. With the aid of this model, information can be processed to be useful for decision-making at enterprise and whole-farming level. These costing procedures taking risk into account have been applied and tested for small- and large-scale commercial farmers in the Nkomazi region of the Komati and Crocodile rivers (Oosthuizen, *et al.* 2005a; 2005b). For small-scale farmers, a subsidy on capital is crucial for financial survival, while for large-scale farmers, the simultaneous replacement of orchards and irrigation equipment has a severe effect on financial feasibility and riskiness of farming. The full financial cost of irrigation, i.e. capital, operating and maintenance cost, was similarly evaluated in a separate case study of smallholder subsistence farming (Perret & Geyser, 2007). The results show high cost of irrigation services in comparison with income from irrigation. This finding supports the approach adopted by DWAF to gradually increase charges for water development and use by subsistence farmers.

4 Hydrological issues

4.1 Metering and complete description of water use entitlements

The security of water use rights implies that it can be monitored and enforced, which further implies that it can be measured. The establishment and enforcement of rights is a central pillar of a market economy and thus a water market. A process is underway to measure or meter water in South Africa. In this regard, it has been shown that technologies are available for direct and indirect measurement of water conveyance in rivers, canals and pipelines. The challenge in practice is managed implementation of the water measuring system, both by individual farmers and by WUAs (Van der Stoep *et al.*, 2005; Du Plessis *et al.*, 2004). Various conditions for licensing of water use, including installation of water measuring devices, are furthermore being implemented by DWAF (Department of Water Affairs and Forestry, 2006a; 2007). As a whole, these actions to measure and monitor should also contribute to a clearer specification and enforcement of water use entitlements.

Alleged illegal use of water is an impediment to a water market especially in two water-stressed areas, namely the Olifants River (East) and the Crocodile River (East) recently visited as part of a Water Research Commission project (Nieuwoudt *et al.*, 2008). In the Olifants River (East), water is metered and monitored in the Loskop Dam and Blyde River irrigation areas and there is no room for illegal use. It is alleged that in areas outside these schemes, the absence of water metering and monitoring is a problem. Mines and farmers are blamed for the alleged illegal abstraction of water. In the water-stressed Crocodile River (East), the lack of metering is alleged to be a problem (Jackson, 2007). Deacon (2005) asserts that some farmers below the Gorge in the Crocodile River (East) have exceeded their entitlements and expanded their acreage. As farmers were allowed to spread their water over a larger area if they use drip or other water conservation measures, it is difficult to enforce entitlements in the absence of metering. Farmers, however, dispute claims of illegal use in the Crocodile (East) River. It is concluded that allocated licences and pumping permits should be enforced.

4.2 Water quality impacts

Not only the volume of water must be conserved but also its quality must be protected. With pollution of water, external costs are generated. The Berg River in the Cape was recently (2007) visited (Nieuwoudt *et al.*, 2008). Stakeholders concerned with the quality have formed an action group in this river. Water quality is also a concern in the Olifants River (East), one of the main rivers in South Africa. Coal mines on this river are allegedly blamed for discharges in the river. The policy recommendations to improve the water quality in this river were highlighted after meeting stakeholders of the Olifants River (East) Forum. Strong support from these stakeholders was received for policy options such as pollution permit trading and environmental offsets.

The catchment surface of the Olifants River (East) is fractured by mining activities, runoff decreases and water is drained into underground aquifers which then seep into streams. Pollution in the river and in Loskop Dam is high and one of the main problems is the effluent leakage from old disused mines. Mines have been permitted to release nutrients in the streams during periods of high flow, which is called the 'controlled release scheme'. During the past few years, river flow was low and sufficient dilution of nutrients was not possible. Mines and power stations had to invest in desalination plants at considerable cost to dispose of pollutants.

A waste discharge charge system is proposed by DWAF (Havenga, 2006), but at present, discharges in the catchment are not levied. It is recommended that polluters should pay a discharge rate, in the same way as water abstraction users pay water charges. As in the case of a water market, it is further proposed that a market be established for the discharge of pollutants and that this market is used to discover the optimum price for pollutant disposal. This proposal is supported by representatives of some mines (Lodewijks, 2006). All markets operate within certain rules. In a pollution permit trading market, rules that may be considered are that discharges in the river are only allowed when flow is sufficiently high and that trades may only occur within certain parameters. A permit-trading programme could complement desalination plants, as some costs of these plants may be variable (reservoirs where the pollutants solidify fill up).

Apart from a pollution-trading programme, it is suggested that biodiversity offsets be created to provide incentives for cooperation amongst stakeholders which may be mines, developers, environmental groups, farmers and public land agencies. Expert opinion is that the main source of pollution in the Loskop Dam is the leakage from abandoned old mines (pre-1956). The problem with the defunct mines is that they leak pollutants all the time including during the period when river flow is low. DWAF has apparently accepted responsibility for these mines but they may not have the appropriate technology, which is also expensive, to desalinate the effluent. In an offsetting arrangement, incentives can be provided to existing mines to desalinate water from these defunct mines by allowing them to discharge a given amount in the Olifants River (East) when the water flow is sufficiently high. The above arrangement will cost the taxpayer nothing while discharge during low flow periods is reduced.

5 Empirical results of water marketing studies

5.1 Efficiency of water use

Studies undertaken in several rivers in South Africa showed that water market trading promotes the more efficient use of water (Gillitt, 2004; Armitage, 1999). Water market trading will promote some of the objectives as stated in the Water Allocation Reform document (Department of Water Affairs and Forestry, 2006b) by supporting growth and development and by maximizing the return, adjusted for risk per unit of water.

A discriminant analysis of water market transfers in the Lower Orange River showed that water rights were transferred to farmers with the highest return per unit of water applied, those producing table grapes, and with a high potential arable 'outer land' without water entitlements (Armitage, 1999). In this analysis, the return per unit of water applied was the most significant of the variables studied and also had the highest standardized regression coefficient. It is concluded that the market promoted more efficient use of water. Buyers of water entitlements only modestly used more water conservation technology as both buyers and sellers face almost the same opportunity cost. The opportunity costs faced by buyers are slightly more because of transaction cost. Only unused water was transferred, while water saved (through adoption of conservation practices) was retained possibly for security purposes.

A second study by Armitage (1999) in the Nkwaleni Valley in northern KwaZulu-Natal found that no water market had emerged despite the scarcity of water in the area. No willing sellers of water rights existed. Transaction costs appear larger than benefits from trading. Farmers generally retain surplus rights as security against drought because of unreliable river flow while crop profitability in this area is similar for buyers and sellers (they grow the same crops). If potential buyers are compared with potential sellers then the most important variable that discriminated between them was that buyers produced sugar cane, the reason being that sugar cane can better withstand drought than competing crops. This may be attributed to the irregular river flow, a finding that was also observed in the Crocodile River (East) (Gillitt, 2004).

A follow-up study by Gillitt (2004) was undertaken among irrigation farmers in the Boegoeberg and Kakamas Irrigation Schemes along the Orange River of South Africa who had transferred water entitlements between January 1998 and August 2003. A total of 37 farmers were interviewed. A principal component analysis of factors associated with buyers in the Orange River was conducted. This indicated that buyers of water entitlements have a higher income per cubic metre of water applied, a larger percentage of cropped area planted to lucrative export table grapes and horticultural crops, and more advanced irrigation technology while it has a negative loadings with the percentage of cropland planted to other grapes and percentage of planted to field crops (lower return). This indicates that a water market promotes efficiency in water use and that water is transferred to high-income crops (table grapes and horticultural crops). These relationships were confirmed in regression models (ridge regression, logistic regression).

5.2 Risk in water marketing

Policy risk and risk aversion appear to be important in explaining future investment in irrigation farming in the Lower Orange River. These farmers are also highly risk averse, especially regarding downside risk. Important policy implications are that farmers should be better informed about the practical implications of the NWA and specifically water licences.

A study was conducted in the Crocodile River (East) where water supply is irregular; water was transferred from areas where risk is high to areas where lower risk crops can be produced especially sugar cane (Gillitt, 2004). A total of 18 farmers were interviewed. Although the number of farmers was small in the latter region, some of these farmers entered into several contracts, for instance, 1 farmer leased from 12 lessors. These conclusions confirm results of a study by Bate *et al.* (1999). In the Lower Crocodile River, crops in the purchasing area have lower production risk (sugar cane), lower financial risk and better cash flow (bananas and sugar cane). Other attributes of buyers are that they irrigate larger areas and they probably have a deficit allocation (use more water than their quota). A motivation for buyers is to legitimize their current use. In regression analyses, the net present value of gross margin per cubic metre of water used was not significant, showing that the main reason for transfer is lower risk.

The characteristics of buyers and sellers of water differ in the Crocodile (East) and Orange rivers. In the Orange River where water supply appears more stable (due to large irrigation dams), water is transferred from areas where the return per cubic metre of water is low to areas where the return is high. In the Lower Crocodile River where water supply is highly irregular, water is transferred from areas where risk is high to areas where lower risk crops such as sugar cane can be produced.

The risk aversion of irrigation farmers was measured by the Arrow–Pratt absolute risk aversion coefficient (APAR) (standardized for scale and range of data) (Nieuwoudt *et al.*, 2005). The empirical investment model shows that farmers who are more risk averse expect to invest less in the future. Risk management strategies in an environment of risk are crucial. Policies that increase risk in agriculture will have a significant negative effect on future investment in irrigation. Farmers were tested for downside risk where there is a chance that they can lose money if they select the uncertain alternative. Farmers are more risk averse (downside risk) than anticipated in the questionnaire as almost all the farmers picked the most risk-averse category. That is, they did not pick a choice where money could be lost. An APAR for downside risk was estimated at 3.28, which implies in a choice situation an indifference between a certain income of R0.0 and being given a 50% chance on winning R800,000 or losing R200,000. The mean of this gamble is R300,000, which is a significant reward for taking a risk.

6 Equity with water allocation

One of the main objectives of the National Water Policy in South Africa is to achieve equitable access to water. A target has been set that at least 30% of water must move towards previously disadvantaged individuals (PDIs) due to racially discriminatory legislation. According to a recent document, it is envisaged that the 30% will be achieved in 2014 while it will be further increased to

45% in 2019 and 60% in 2024. Of the 60% attributable to Blacks in 2024, it is suggested that half should be allocated to women (Department of Water Affairs and Forestry, 2008). If irrigation water is provided to PDIs then they will still have to be provided with suitable irrigable land, which is a problem as most of the suitable land is already under cultivation. It does not make sense if water is moved from a developed farm which has little production potential without water and channelling it to a new farm that must still be developed. The development cost of irrigation farming is high and providing PDIs with water only does not make sense. It is proposed that the most effective way to redistribute water to PDIs is through the government programmes of restitution and redistribution of land. The value of water is capitalized in the value of a farm and empowering a PDI to own a farm also provides him access to the water rights of the farm. PDIs need not only water and land but also other support services to build their capacity, which includes training. This implies coordination of actions from the government departments of land affairs, water affairs and forestry, and agriculture (extension service) while other stakeholders need to be included such as the Land Bank of South Africa (financing) and commercial farmers.

In two of the areas visited (Crocodile River (East) and Olifants River (East)) land claims under restitution exceeded the targeted 30%. According to Van Veyeren (2007), approximately 80% of the land in the Crocodile River (East) has already been sold to PDIs, which means that 80% of the water rights have moved to PDIs. In the Berg River, which was also visited during 2007, no land claims exist. The government programme of land redistribution, however, allows for PDIs to purchase land by using grants from the Department of Land Affairs and coupling it with loans from the Development Bank of South Africa. It is suggested that the latter programme be considered for this area.

7 Governance in water management

Water governance refers to the range of political, social, economic and administrative systems that are in place to develop and manage water resources, and the delivery of water services (Rogers & Hall as quoted by Pegram *et al.* 2006). ‘Good governance’ depends on the principles of predictability, inclusivity, representivity, accountability, efficiency, effectiveness, social equity and justice. Other principles such as transparency are necessary to ensure safeguards in the system, while cooperation is necessary in a highly complex system. Causes of ‘ineffective governance’ include corruption, inadequate financial resources, inadequate labour and managerial skills, low prioritization and poor communication (Moss as quoted by Pegram *et al.* 2006).

According to Pegram *et al.* (2006), ‘coherent governance of the water environment has been simplified by the definition of water resources under the NWA to include a “watercourse, surface water, estuary, or aquifer”, … while resource quality refers to the quantity, quality, habitat and biota of a water resource. These broad definitions, together with the broad definition of water use to include abstraction, storage, streamflow reduction, waste discharge (including sea outfalls), waste disposal (with impact on water), in-stream activities and recreation, provide a relatively integrated basis for water resources management.

South Africa's water resources policy and legislation is firmly grounded in the principles of the Constitution, and considers international best practice around integrated water resources management (IWRM) (including decentralization and participation) within the historical context of South Africa requiring redress. The NWA develops a coherent and integrated governance framework around these concepts, addressing catchment level strategic planning, allocation, protection, development and utilization of the water resources and charging for water. It further provides for decentralized organizational framework for water resource management, based on the establishment of CMAs.

However, the development and implementation of the necessary regulatory enabling framework for legislation and regulation has been slow, particularly in the delayed establishment of CMAs; transformation of WUAs; establishment of a water resources classification system; reallocation of water use entitlements, including compulsory licensing; development of catchment management strategies; authorization of water use within a catchment paradigm; development of economic instruments under the charging strategy, such as the waste discharge charge system.'

CMAs are statutory bodies established in terms of Chapter 7 of the NWA for the management for water resources. CMAs are responsible for the planning, implementation and management of water resources. Second, they are established to coordinate the water-related activities of other organizations and water users. CMAs also play a role as organizations to which certain functions currently performed at national level may be delegated at regional or catchment level. The initial functions of the CMAs are mainly centred on managing the regional water resources and ensuring stakeholder participation within a water management area (WMA).

WUAs are cooperative associations of water users established under the NWA to undertake water-related activities for the mutual benefit of all its members within a WMA. Within the associations, members cooperate and pool resources to address local water-related needs and priorities. WUAs are therefore mainly established to manage local water infrastructure, e.g. irrigation water supply schemes, and to implement management decisions agreed on between the members (Mazibuko & Pegram, 2006).

The Inkomati CMA establishment process began in 2000 and it was finally launched (the first in South Africa) in 2006. The Inkomati WMA is a combination of the Komati, Crocodile and Sabie-Sand catchments. Within its jurisdiction falls the Crocodile (East) Main Irrigation Board which still has to be transformed to a WUA. At present, commercial agriculture, with crops such as sugar cane, citrus and other sub-tropical fruits, is the main water user in the catchment (Van Vuuren, 2007).

Pegram *et al.* (2006) further state that "delays in the development of key regulatory instruments have meant that the institutional and practical implementation of the policy and legal framework for water governance is not well developed. While this is not ideal, it has allowed improved understanding and implicit change within the sector and implies that the fundamental regulatory change may be introduced in a coherent manner over the next few years. Specific governance issues related to the water environment include

- Governance of CMAs at all levels, within a paradigm of cooperative, integrated, developmental and participatory management:
  - Institutionalizing stakeholder participation in decision-making
  - Oversight responsibilities, instruments and capacity of DWAF to ensure effective governance and functioning of CMAs
- Governance considerations and mechanisms for the development of catchment management strategies through a consultative process and their alignment with local development planning and provincial planning processes
- The institutional change (at all levels) required to move from centralized regulatory DWAF decision-making to decentralized participatory governance in water resources management
- Requirements to empower civil society (particularly rural communities, women and the poor) to promote effective water governance at a local catchment scale".

8 Separation of water and land rents

Research initiated and financed by the Water Research Commission of South Africa in the period 2005–08 (Nieuwoudt *et al.*, 2008) in three catchments concluded that through the water market, land and water prices will be further separated as water moves to non-agriculture with the concomitant increase in water prices and a decrease in land prices. The reason is that some of the investments in irrigation such as development costs of land are fixed (sunk) with zero opportunity costs. Without water, the development cost of land has no value in its second best use (grazing). Water is expected to move to urban use because of the high migration in South Africa to cities. Water prices may increase as follows: in the Crocodile River (East) from R15,000 ($1=R10) ha^{-1} (13,000m^3 ha^{-1}) to R45,000 ha^{-1} (or R3.46 m^3), in the Olifants River (Loskop Dam) from R11,673 ha^{-1} (7700m^3 ha^{-1}) to R19,500 ha^{-1} (or R2.53 m^3), while in the Berg River from R20,000 ha^{-1} (5000m^3 ha^{-1}) to R45,000 ha^{-1} (R9 m^3). The purpose was not to calculate a precise figure for future water prices but to show that in a water market, prices will increase in future providing an incentive for further water conservation and more efficient use. The latter is important as the increase in food prices experienced during 2008 may be a long-run phenomenon. This agrees with the view of the South African National Agricultural Marketing Council (2008) that governments and international organizations have notoriously underinvested in food, nutrition and agriculture.

9 Conclusion

Considerable progress has been made with implementation of new approaches to water management according to the legal framework in South Africa. It is anticipated that acceptance of regulations to mandate water measurement will enable volumetric charging. This is an important economic instrument to relate financial water costs to water use and to provide incentives for efficient irrigation. At the same time, a change from a unitary to a two-part charging system should be introduced (Water Research Commission, 1996). It will facilitate both demand management of water and balancing of the budgets of CMAs and WUAs. Correct incentives for conservation and allocation of water will be instituted by a combination of accelerated compulsory licensing to achieve equity and promotion of water markets to achieve efficiency. Key requirements are restitution or redistribution of land together with water use rights and clear specification of the volume and reliability of available water attached to the use entitlement. Decentralization of water management functions from DWAF central offices to CMAs finally requires cooperation and participation by all water users to ensure effective performance and governance.

References

Armitage, R.M. 1999. An economic analysis of surface irrigation water rights transfers in selected areas of South Africa. WRC Report No 870/1/99, Pretoria, South Africa.

Bate, R., Tren R. & Mooney L. 1999. An econometric and institutional economic analysis of water use in the Crocodile River Catchment, Mpumalanga Province, South Africa. WRC Report No 855/1/99, Pretoria, South Africa.

Backeberg, G.R. 2005. Water user charges in irrigated agriculture: Principles and practice. Combined Congress of the South African Societies for Soil Science, Crop Production, Weed Science and Horticultural Science, 10–13 January 2005, Potchefstroom, South Africa.

Backeberg, G.R. 2007. *Allocation of Water Use Rights in Irrigated Agriculture: Experience with Designing Institutions and Facilitating Market Processes in South Africa.* USCID 4th International Conference on Irrigation and Drainage, 3–5 October 2007, Sacramento, California.

Carruthers, I. & Clark, C. 1981. *The Economics of Irrigation.* Liverpool University Press: Liverpool.

Deacon, E. 2005. Personal Communication, DWAF, Nelspruit, South Africa.

Department of Water Affairs and Forestry. 2006a. Water conservation and water demand management conditions: use of water for domestic, agriculture and industrial sectors (Final report version 1.0). Directorate: Water Use Efficiency, Pretoria, South Africa.

Department of Water Affairs and Forestry. 2006b. Guidelines for water allocation in South Africa: A framework for water allocation reform (Draft 2.2). Directorate: Water Allocation, Pretoria, South Africa.

Department of Water Affairs and Forestry. 2007. Regulations limiting or restricting the purpose, manner or extent of water use, requiring that the use of water from a water resource be monitored, measured and recorded: regulating the design, construction, installation, operation and maintenance of any waterwork and regulating or prohibiting any activity in order to protect a water resource, Draft 3 for comment June 2007. Directorate: Water Use Efficiency, Pretoria, South Africa.

Department of Water Affairs and Forestry. 2008. *Water Allocation Reform Strategy.* Chief Directorate: Water Use, Pretoria, South Africa.

Du Plessis, F.J. 2004. Electric power supply measurement as an alternative to measure flow-rates of hydraulic pumps. WRC Report No 1190/1/04, Pretoria, South Africa.

Genesis Analytics. 2005. Part 1 The Contribution, Costs and Development Opportunities of Forestry, Timber, Pulp and Paper Industries in South Africa, Private Bag X1, Melrose Arch, Johannesburg, www.forestry.co.za accessed June 2007.

Gildenhuys, J.S.H. 1997. *Public Financial Management*. J L Van Schaik Publishers, Pretoria, South Africa.

Gillitt, C.G. 2004. Water markets in irrigation areas of the lower Orange and Crocodile rivers of South Africa. WRC Report No KV 160/04, Pretoria, South Africa.

Gittinger, J.P. 1982. *Economic Analysis of Agricultural Projects* (2nd edition). The Johns Hopkins University Press, Baltimore and London.

Havenga, B. 2006. Personal Communication, DWAF, Pretoria, South Africa.

Hobbs G.J. Jr., Hobbs, T. & Raley, P.C. 1994. Institutional roles and water marketing in Colorado and the Western United States. Paper presented to World Bank Group, Water Legislation, An integrated approach, September 8–9, Washington, DC.

Jackson, B. 2007. Personal Comment, DWAF Office, Nelspruit, South Africa.

Lodewijks, H. 2006. Personal Comment, Anglo Coal, South Africa.

Meiring, J.A., Oosthuizen, L.K., Botha, P.W. & Crow, C.I. 2002. The implementation of the FARMS system for management decision-making in irrigation farming: Summary reports and user guides. WRC Report No 894/1-4/02, Pretoria, South Africa.

National Water Act (NWA). 1998. No 36 of 1998 as amended, Republic of South Africa, Pretoria, SA, available at: http://www.dwaf.gov.za, 1998.

National Agricultural Marketing Council. 2008. *The South African Food Cost Review: 2007*. NAMC and the Department of Agriculture, Pretoria, South Africa.

Nieuwoudt, W.L., Gillitt, C.G. & Backeberg, G.R. 2005. Implication of risk in irrigation water transfers in South Africa, Proceedings of International Conference on Water Economics, Statistics and Finance, Ed. KP, 2005.

Nieuwoudt, WL, Döckel, JA, Mosaka, D & Pott, AJ. 2008. Towards the establishment of water market institutions for effective and efficient water allocation in South Africa. WRC Report No. 1569/1/08, Pretoria, South Africa.

NWRS, 2004. National Water Resource Strategy, Department of Water Affairs and Forestry, Pretoria, South Africa.

Mazibuko, G. & Pegram, G. Evaluation of the opportunities for cooperative governance between catchment management agencies and local government. WRC Report No 1433/1/06, Pretoria, South Africa.

Oosthuizen, L.K., Botha, P.W., Grové, B. & Meiring, J.A. 2005a. Cost-estimating procedures for drip-, micro, and furrow-irrigation systems. *Water SA*, 31(3), pp. 403–406.

Oosthuizen, L.K., Botha, P.W., Grové, B., Meiring, J.A., Monkhei, M.M. & Pretorius, I.G. 2005b. Cost Estimating Procedures for micro-, drip- and furrow-irrigation systems for large- and small-scale farmers in the Onderberg/Nkomazi region. WRC Report No 974/1/05, Pretoria, South Africa.

Pegram, G., Mazibuko, G. Hollingworth, B. & Anderson, E. 2006. Strategic review of current and emerging governance systems related to water in the environment in South Africa. WRC Report No 1514/1/06, Pretoria, South Africa.

Perret, S. & Geyser, M. 2007. The full financial costs of irrigation services: A discussion on existing guidelines and implications for smallholder irrigation in South Africa. *Water SA*, 33(1), pp. 67–78.

Van der Stoep, I., Benadé, N., Smal, H.S. & Reinders, F.B. 2005. Guidelines for irrigation water measurement in practice. WRC Report No TT 248/05, Pretoria, South Africa.

Van Veyeren, C.G. 2007. Personal Comment, Vice Chairman Crocodile River Main Irrigation Board, Nelspruit, South Africa.

Van Vuuren, L. 2007. Joining hands to manage SA's water resources. *The Water Wheel,* July/August 2007.

Water Research Commission. 1996. Policy proposal for irrigated agriculture in South Africa. Discussion Paper prepared by Backeberg, G.R., Bembridge, T.J., Bennie, A.T.P., Groenewald, J.A., Hammes, P.S., Pullen, R.A. & Thompson, H., WRC Report No KV96/96, Pretoria, July 1996.

Chapter 14

Sustainability of groundwater resources in India: Challenges and scope for economic instruments and policy

S. Msangi
International Food Policy Research Institute, Washington DC, USA

Abstract

In this chapter, we address the groundwater resource management issues in selected regions of India, within the context of the policy and natural environments that exist. We illustrate the underlying nature of the resource management problem and suggest possible alternative interventions, with the use of an economic model that is linked to a simplified representation of the characteristic hydrology. By illustrating the policy problem in this way, the chapter explores the scope that alternative economic instruments could have in correcting the perverse incentives that exist for groundwater conservation, in some regions, while improving human welfare. In doing so, we hope to better clarify the role of market (and non-market based) instruments in addressing common pool resource management problems in India, and provide guidance to researchers and policymakers on how they can best study these cases, and further refine their policy recommendations.

Keywords: Common pool resources; Groundwater; Market-based instruments; Natural resource management; Policy and institutions

1 Introduction

The critical situation of groundwater overdraft in South Asia has been under keen observation in the last decade, given the increasing resource scarcity that has been felt by rural farmers in that region, especially in India. Despite the tremendous productivity gains to Indian agriculture, which were realized during the Green Revolution, the problem of resource scarcity has been aggravated by the introduction of these water-intensive varieties to an extent where current levels of groundwater extraction are unsustainable. Besides the scarcity of land in India, there has been a keen competition for limited water resources on which many of the important cereals, pulses and even higher value crops depend for continued productivity in the face of variable environmental and economic conditions. The tension between larger landowners and those that are less well-endowed often stems from the differential access to key productive inputs such as water. As has been observed in the North China Plain, decades after the policy-induced explosion in groundwater exploitation to promote grain security (Crook, 1999), the rapid increase in tubewell usage, and the expansion of groundwater irrigation in India, has also had its toll on the natural resource base, over time.

The literature on water resources in India has documented the problem of groundwater overdraft and identified a number of issues to be at the root of the problem. An impressive example has been set by a handful of charismatic village leaders in Western India who are able to mobilize awareness over economic and environmental issues, as well as exert their personal influence over villagers to change both their social and resource usage behaviour. Many policymakers and analysts are intrigued by their example, and seek to better understand the elements of their success – and how it might be replicated in other settings. As it is difficult to 'replicate' such leaders – well-designed policy mechanisms must take their place, to help coordinate the activities of farmers and overcome the cooperative failures that might otherwise result in decreased overall benefits.

In this chapter, we point to a number of promising ideas that policymakers can apply to the problem of managing the demand for scarce groundwater resources, since options for supply expansion are largely exhausted. There are a number of important institutional considerations that we point out, which are critical in determining the success of these policy measures. We illustrate where market-based instruments might be appropriate, and where a more centralized form of intervention might be needed to stabilize the resource base.

In the next section, we present a quick overview of the issues facing groundwater users and resource managers in India, which is followed by a discussion of the economics underlying the natural resources management problem. The subsequent section highlights some possible types of intervention, and a concluding section summarizes the main arguments and offers directions for future study.

2 Overview of groundwater issues in India

The importance of irrigation in maintaining the necessary productivity within the Indian agricultural sector has long been recognized by researchers and policymakers, and is a major driver behind the growth in output that was observed during the period of the 'Green Revolution'. While the agro-ecological conditions vary widely across the Indian subcontinent, there is a sizable share of agricultural production that relies on irrigation, as shown in Table 1.

Much of the literature that deals with the management of water resources in Indian agriculture has focused, in particular, on groundwater management, which represents a significant share of the water resources that are available to the

Table 1: India irrigation, 2003–04.

State	Total cultivable area ('000 ha)	Net irrigated area ('000 ha)	Irrigated as percentage of total cultivable (%)
East	13,077	2416	18.5
West	76,758	16,862	22.0
North Central	36,529	18,977	52.0
North East	6129	439	7.2
North West	11,477	7818	68.1
South	39,498	8592	21.8
All India total	183,468	55,104	30.0

Source: CWC, Delhi, India – Water-related statistics.

Table 2: Indian irrigation by source.

State	Tubewells ('000 ha)	Other wells ('000 ha)	Other sources ('000 ha)	Total all sources ('000 ha)
East	233	98	265	2416
West	3865	8220	893	16,867
North Central	12,898	569	622	18,977
North East	2	2	193	437
North West	4865	20	183	7818
South	2304	2186	597	8592
All India total	24,167	11,095	2753	55,107

Source: CWC, Delhi, India – Water-related statistics.

majority of Indian farmers. Table 2 shows the agricultural area that is supported by tubewells and other wells, as opposed to other kinds of water withdrawals from surface sources.

The deepening scarcity problems that have been observed in many parts of India, as a result of increasing demands on limited water resources for both agricultural and non-agricultural uses, have also been a source of concern for both policymakers and analysts who seek to promote the sustainability of water resources through improved management on both the demand and the supply side. A number of authors have focused on the electricity subsidies to agriculture, which might be giving perverse incentives to farmers using electric pumps to withdraw more groundwater than would otherwise be socially optimal (Kumar, 2005; Scott & Shah, 2004). The lowering of electricity costs, among other subsidies to the agricultural sector (e.g. fertilizer) that have been ongoing since the 1970s, have been an integral part of the strategy of the government to encourage the intensification of agriculture. This has led to flat electricity tariffs and radically reduced costs of power, resulting in the virtual demetering of rural zones. Others have suggested alternative policy instruments that might help groundwater users to internalize the external effects that they impose on other users of the common pool groundwater resource, through such instruments as taxes on pumping, based either on the volumetric quantity of water used or electricity that is consumed (Kumar & Singh, 2001).

These types of instruments, however, are difficult to impose, due to problems of observability and measurability of actual volumes of water that are used by individual groundwater users. The recommendations that might apply to regions that have groundwater pumpers with very large landholdings, like Kern County, California (Feinerman & Knapp, 1983), may not be as applicable to developing country regions where there are numerous smallholder farmers, like in East and South Asia. A number of other studies have tried to examine the role of markets for water, and the way they can potentially improve the allocation of water among users, by using market-based incentives and trading mechanisms, as well as the possible welfare losses that can result from the exercise of monopoly power by wealthier and larger landowners on their poorer and smaller neighbours (Kajisa & Sakurai, 2005; Prakash, 2005). Nonetheless, the presence of informal markets for water has been shown to provide benefits for smaller landowners or tenant farmers who would otherwise not have the means to make the investment in tubewells themselves (Meinzen-Dick, 1996).

In this chapter, however, we will focus on the choice of instruments that might be employed by policymakers to encourage more efficient use of water resources, and promote the sustainability of limited groundwater resources, within the context of India. We will consider the efficacy of tax- or tariff-based approaches to control, as well as that of quantity-focused instruments – and discuss the situations in which the efficacy of one might outweigh that of the alternative methods. We will discuss these instruments within the context of a theoretical model of economic behaviour, which addresses groundwater use as well as inter-agent reallocation and trade of permits. Our conclusions and recommendations will then close the chapter.

3 Linkages between land, energy and water usage

We first examine the linkages that exist between land and water usage, before considering the role that energy plays in affecting groundwater usage.

3.1 The economics of water usage and the role of land

The economics of water usage is typically based on the behavioural economics of a profit-maximizing agent, who seeks to maximize the net revenue that accrues from irrigated agricultural production, and faces a trade-off in terms of costs of inputs (including water), or constraints in water or land use. Even without considering the dynamics of groundwater usage, and how the stock of water held in the underground aquifer evolves over time, the individual profit-seeking agent can be hypothesized to behave according to the following, simple maximization problem.

$$\max_{x,A} \quad pAf(x) - cx - \alpha A^2 \qquad s.t. \quad A \leq \bar{A} \tag{1}$$

where the decision-maker's problem is defined in terms of choosing the optimal level of input (i.e. water) which enters into the agricultural yield function $f(x)$, while also choosing the optimal land area over which to farm, A, which is available up to a particular limit $\bar{A}$. The price of the agricultural output is denoted by p, whereas the cost of the productive input is given as c. The quadratic term in land area, α, captures the decreasing returns to adding land area which is due to limited management and labour, as well as variable land quality over the available area. The maximization problem written in eqn (1) can be stated in terms of the full Lagrangian function, shown in eqn (2) below, which has the shadow value of the constraint included as a choice variable.

$$\max_{x,A,\lambda} \quad L(x, A, \lambda) = pAf(x) - cx - \alpha A^2 - \lambda\left[A - \bar{A}\right] \tag{2}$$

The necessary conditions for profit-maximizing choice over the variables x and A lead to the following set of equations

$$\begin{array}{lll} pAf'(x) - c \leq 0 & x \geq 0 & x\left[pAf'(x) - c\right] = 0 \\ pf(x) - 2\alpha A - \lambda \leq 0 & A \geq 0 & A\left[pf(x) - 2\alpha A - \lambda\right] = 0 \\ A - \bar{A} \leq 0 & \lambda \geq 0 & \lambda\left[A - \bar{A}\right] = 0 \end{array} \tag{3}$$

which we can use to define the optimal choice of water and land usage by the representative farmer. If we were to assume that the decision-maker always chooses to use as much land as is available, then we can reduce the set of conditions in eqn (3) to the following pair of equations

$$\begin{array}{l} p\bar{A}f'(x) - c = 0 \\ pf(x) - 2\alpha\bar{A} - \lambda = 0 \end{array} \tag{4}$$

which we can use to examine the sensitivity of the optimal choices to the key economic parameter values. By totally differentiating the pair of first-order conditions in eqn (3), with respect to each variable and parameter, we obtain the following linear system

$$\begin{bmatrix} p\bar{A}f''(x) & 0 \\ pf'(x) & -1 \end{bmatrix}\begin{bmatrix} dx \\ d\lambda \end{bmatrix} = \begin{bmatrix} -\bar{A}f'(x) & -pf'(x) & +1 \\ -f(x) & +2\alpha & 0 \end{bmatrix}\begin{bmatrix} dp \\ d\bar{A} \\ dc \end{bmatrix} \tag{5}$$

which allows us to relate changes in the vector of decision variables, on the left-hand side, to the vector of the key parameters, on the right-hand side. The sign of the principle determinant, *D*, shown below

$$D = \begin{vmatrix} p\bar{A}f''(x) & 0 \\ pf'(x) & -1 \end{vmatrix} > 0 \tag{6}$$

conforms to our expectations of a well-behaved maximization problem, in which the set of production possibilities can be circumscribed by a convex hull. Applying Cramer's rule to the linear system in eqn (5) allows us to perform the following comparative static calculations, in which we examine the impact of parameter changes on the key decision variable of interest – that of water use (x). To look at the impact of changes in the volumetric cost of water (c) on water usage, we can compute the following differential

$$\frac{\partial x}{\partial c} = \frac{\begin{vmatrix} +1 & 0 \\ 0 & -1 \end{vmatrix}}{D} = \frac{-1}{D} < 0 \tag{7}$$

in which the marginal effect of increasing the cost serves to decrease the level of water usage, as we would expect. Similarly, we can examine the impact of changing the constraint on land, such that we allow for a marginal change in the binding quantity $\bar{A}$.

$$\frac{\partial x}{\partial \bar{A}} = \frac{\begin{vmatrix} -pf'(x) & 0 \\ +2\alpha & -1 \end{vmatrix}}{D} = \frac{pf'(x)}{D} > 0 \tag{8}$$

While these effects are opposite in sign, we can see that their relative magnitude depends on the magnitude of the value marginal product of water $pf'(x)$ relative to unity. If it turns out that the value marginal product exceeds unity, then a change in allowable land area might be more effective in reducing water usage, than a change in the unit variable cost of water. The converse would be implied by the case where $pf'(x) < 1$. While the comparison of marginal impacts on water use, from the point of view of comparative statics, is useful in judging the efficacy of alternative instruments for managing water demand, it does not describe the relative efficacy of the institutions that might be used to apply those instruments. We will now take up this discussion in the following section.

3.2 The energy dimension of groundwater usage

Energy enters the groundwater user's problem, when consideration is given to the cost of pumping water to the land surface from the groundwater table. The unit cost of water usage, which we described by the single parameter c, in eqns (1) and (2), can be conceptualized as a function of several variables, within the context of groundwater usage. Typically, the marginal cost of groundwater pumping is thought to vary according to the hydrological conditions under which water is withdrawn from the aquifer. In particular, the distance over which water must be lifted from the groundwater table to the surface (i.e. the 'lift') is a key determinant to the marginal cost of pumping a single volumetric unit of water, as well as the energy costs that are associated with the action of the pump. Therefore, we can describe the marginal cost of water as a function of the 'state' of the system, which we can describe by the state variable h, which denotes the height of the groundwater table with respect to a reference level, and the distance it lies below the ground surface , shown in Figure 1 below.

Combining the 'lift' (3.) with the energy cost of pumping (e), we can express the marginal cost of water usage as

$$c(h,e) = \gamma e \cdot \left(\bar{S} - h\right) \tag{9}$$

where γ is a conversion factor. Combining this expression for marginal cost with the expression of the farmer's first-order condition for water usage in eqn 4 leads to

$$p\bar{A}f'(x) = \gamma e \cdot \left(\bar{S} - h(x_{-t})\right) \tag{10}$$

in which we have also expressed the groundwater table, as a function of water usage decisions in the past x_{-t} and which relates the present state-of-natural to a history of resource usage behaviour which, over time, results in increasing levels of cost to the farmer, both in the immediate period and in the longer term. In the next section, we look at the way in which a more dynamic perspective can lead to more efficient usage of water resource by agricultural users.

4 Alternative economic instruments for Indian groundwater

Now that we have discussed the technical aspects of groundwater user behaviour, with respect to key parameters of economic behaviour, we can turn to the institutional aspects of implementing groundwater management policy.

4.1 Demand-side management through quotas

If we were to introduce a volumetric charge for pumping, then this would act as an additive factor that raises the marginal cost of pumping above that which is

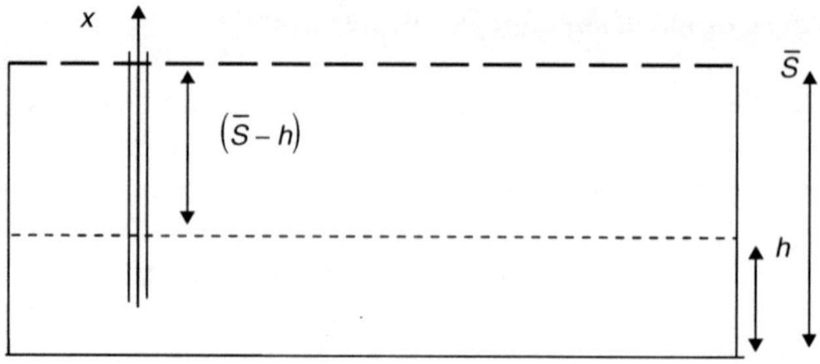

Figure 1: Simplified representation of pumping from an aquifer.

determined by energy and the hydrological state of the system. Such a charge or tax, t, would change the marginal cost in eqn (9) to equal

$$\tilde{c}(h,e,t) = \gamma e \cdot \left(\bar{S} - h\right) + t \tag{11}$$

If we were to use the upper limit on available land as a policy instrument that limits the area of a heavily water-consumptive crop, we could then specify a limit that either coincides with or falls below the 'natural' limit that the farmer would otherwise face in the absence of policy intervention. If we denote this limit as a quota ($\bar{Q}$), which would be allocated to each irrigator then we would have

$$\max_{x,A} \quad pAf(x) - c(h,e,t)x - \alpha A^2 \qquad s.t. \quad A \leq \bar{Q} \leq \bar{A} \tag{12}$$

which represents the modified maximization problem of the irrigator who faces two possible policy instruments. Applying the same assumptions as those used to derive eqn (4), we would have a modified set of first-order necessary conditions

$$\begin{aligned} p\bar{Q}f'(x) - \gamma e \cdot \left(\bar{S} - h\right) + t = 0 \\ pf(x) - 2\alpha\bar{Q} - \lambda = 0 \end{aligned} \tag{13}$$

which yields a similar linear equation system when we take the total differentials with respect to all the decision variables and parameters of the problem. The magnitude of the differential, with respect to the tax, is identical to that shown in eqn (7), and is only a function of the parameters in the determinant. We can derive an additional expression that illustrates the impact of the quota on the implicit shadow value λ, which can be written as

$$\frac{\partial \lambda}{\partial \bar{Q}} = \frac{2\alpha p\bar{Q}f''(x) + \left(\tilde{c}/\bar{Q}\right)^2}{D} < 0 \tag{14}$$

and is negative in sign, if the marginal pumping cost is sufficiently small compared to the allocated quota. The sign of this expression conforms to what we would expect from a downward sloping demand curve for allocated quota, and denotes that the willingness to pay for additional quota goes up as the allocation gets smaller.

In the hands of a well-informed agent, who is forward-looking with respect to the state of the groundwater table, and the implications that a lower water table has for the future pumping costs and producer welfare, we could construct an optimal path of extraction that would ensure the long-term efficiency of water usage, and maximize the sustainability of groundwater usage. Such a path would be obtained by solving the following social planner's problem.

$$V^{CP}(h) = \max_{x,A} \left\{ \begin{array}{l} pAf(x) - \gamma e \cdot (\bar{S} - h)x - \alpha A^2 + \beta V^{CP}(h + \varphi x - \tilde{r}) \\ s.t. \quad A \leq \bar{A} \end{array} \right\} \quad (15)$$

where the inter-temporal optimization is carried out with respect to the pumping of all players in each period, and where β is the discount rate that captures the social planners inter-temporal preferences. The function $V(h)$ is the maximized value of the dynamic game problem, for each player, beginning with the current level of groundwater lift (h), and proceeding under the assumption that actions taken in subsequent periods are done optimally with respect to the groundwater lift in each period. This recursive relationship linking the implied optimality of behaviour from period to period captures the essence of Bellman's 'Principle of Optimality' (Bellman, 1957).

The solution to the social planner's problem would give a 'benchmark' outcome for water and land use that would maximize long-run benefits over time, and enhance groundwater resource sustainability over time. This benchmark would guide the policymaker as to the best choice of tax ($\hat{t}$) or quota ($\hat{Q}$) to impose as an economic instrument on the less-informed, myopic irrigator who would not otherwise consider the long-run benefit. Where numerous irrigators are involved, the social planners' problem can be generalized to yield a vector of decisions over all agents that prescribe their optimal pumping $\{\hat{x}_i\}_{i=1}^{N}$ and land use $\{\hat{A}_i\}_{i=1}^{N}$ behaviour, over time. The well-informed policymaker would choose the optimal quota allocations as $\hat{Q}_i = \hat{A}_i$ and the optimal pumping tax ($\hat{t}$) equal to $\beta V'(h + \varphi x - \tilde{r})$, which represents the extra marginal cost that the social planner imputes to each unit of water withdrawn from the aquifer, and is proportional to the derivative of the optimal value function.

The information that would be needed by the policymaker to set these policy instruments to their optimal levels is daunting, when dealing with many individual agents, and is beyond the capacity of typical institutions of groundwater management. The fact that the individual irrigators know their own technologies and productive possibilities much better than the central administrator is a persuasive argument behind using market-based instruments – such that the individuals can interact within a decentralized framework, on the basis of their privately held information.

4.2 Market-based instruments for demand allocation

Allowing the quotas for land area to be tradable allows each player to engage in decentralized transactions that permit them to trade their initial allocations up to the point that their private benefits are maximized, and their individual shadow values for quota allocations (λ_i) are equalized across agents. Taking the case of just two agents into consideration, for simplicity, we can depict such an equilibrium outcome in terms of Figure 2. This figure shows the equilibrium outcome where transaction costs exist (τ), which cause the agent transactions ($\hat{z}$) to deviate from those levels that would be realized in the absence of transaction costs (z^*). In the presence of transactions costs, the volume of trade is less than that which is otherwise achievable, and the aggregate benefits of both agents fall below that which the central planner could achieve in the benchmark case.

4.3 Payment for environmental services

Borrowing from the notion of paying agents to perform a valuable environmental service such as saving water, we perform a simple experiment to try and quantify what such type of a payment would translate to in the context of rural Maharashtra state in central India.

Drawing on data from two villages in Ahmednagar district of Maharashtra (Hivre Bazaar and Shivni), we construct a numerical model that simulates water usage behaviour in agriculture and the subsequent effects on groundwater levels. We use this framework to measure the equivalent payment that would need to be made to farmers, in order for them to forego higher levels of groundwater pumping in the initial period, in return for lower pumping costs in the future. In other words, we contrast the rapid drawdown of the water table in the early periods, with a more gradual path of water abstraction that enables users to withdraw water at a continuous and steady rate into the future, thereby avoiding high levels of pumping costs, corresponding to the higher levels of groundwater lift that they would otherwise be facing in the future.

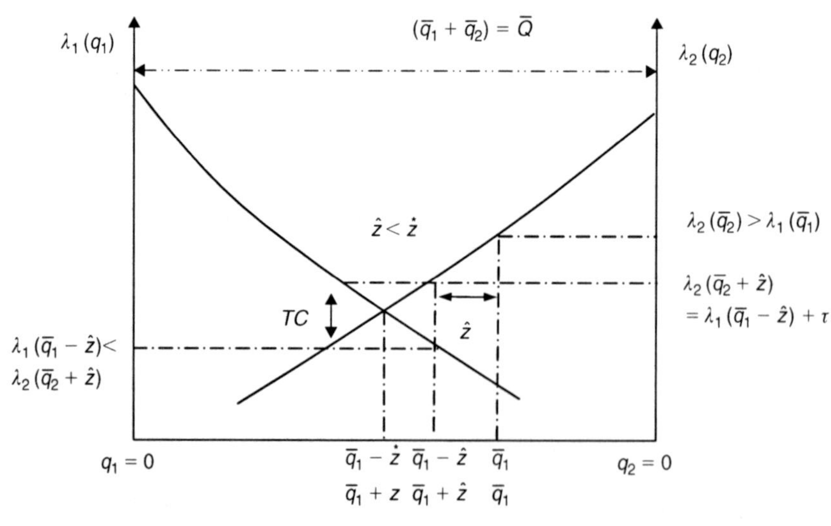

Figure 2: Decentralized allocation of tradable quota between agents.

The more gradual path of extraction, which we describe above, emanates from the forward-looking, dynamic behaviour of an idealized type of user who recognizes the long-term benefits of keeping water in the ground and appropriately balances this with the short-term benefits of extraction. This type of 'social planner' model gives a prescriptive view to the way in which the aquifer *should* be managed, and should be viewed as a way of benchmarking groundwater efficiency. In essence, the social planner's optimization problem solves the dynamic resource extraction problem, in a way that satisfies the following relationship

$$\begin{matrix}\text{Maximized} \\ \text{present value}\end{matrix} = \max_{\text{decision variables}} \left\{ \begin{matrix}\text{current benefits} + \beta \cdot (\text{maximized} \\ \text{future value})\end{matrix} \right\}$$

where the decision variables, namely the rate of groundwater withdrawals, drives the state of the system and determines the short-term gain while also affecting the future gains that can be realized from the water remaining in future periods.

The comparison between the optimal solution of the social planner and that of the myopically extracting case actually provides this type of information, since we can see exactly how the initial benefits gained by extracting more groundwater than is optimal compare with the later additional costs that are incurred in pumping from a lower groundwater table.

In the case of Hivre Bazaar village, we can see the comparison between the stream of net benefits that are derived from groundwater withdrawals over time – from both the optimal case and that of myopic extraction (Figure 3).

From this comparison, we can see that the gain in net benefit that the myopic extraction case enjoys in the initial period is quickly dissipated, as the ground

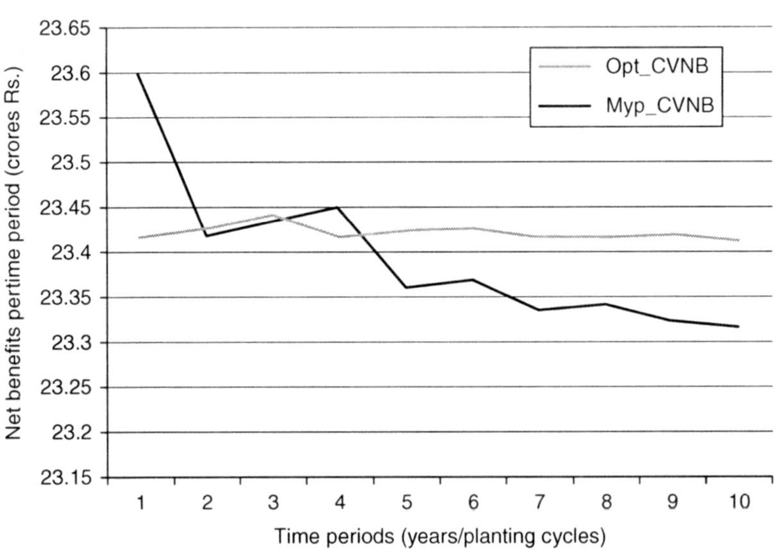

Figure 3: Comparison of current value net benefits under optimal and sub-optimal extraction (Hivre Bazaar village).

water table drops, and higher pumping costs are incurred. The net benefits from the 5th year onwards remain consistently (and increasingly) below those of the social planner's optimal path, which remains fairly constant over time. For the purposes of comparison, we can contrast the initial net gain of myopic pumping – of roughly Rs 180,000 total – to the subsequent losses, which must be converted to a net present value, in order to remain comparable with the initial period gain. In this case, we apply the same 5% discount rate that is used in the model simulations. This gives us a figure of Rs 11,849 per household in initial gain to myopic behaviour – compared to the net present value of Rs 21,390 per household in loss that is incurred because of higher pumping costs in the subsequent time periods. This means that, in net present value terms, losses more than double the initial gains that are made by over-pumping. A 'patient' household would be able to realize the future gains if they were sufficiently forward-looking; however, this is not the case for myopically behaving agents who seek to maximize only their current profits, without regard for the carry-over effects.

A similar outcome is observed for Shivni village, in Figure 4, and the same initial gain is soon followed by subsequent losses in net benefit over time, as the groundwater table deepens. Even though the overall magnitude of gains and losses differ from those of Hivre Bazaar, the overall patterns and implications are the same.

In the case of Shivni, the initial gain due to myopic over-extraction (of Rs 220,000 total) translates to roughly Rs 5007 per household, and is balanced by a loss over the subsequent period equal to the net present value sum of Rs 12,845, which is more than double the initial gain.

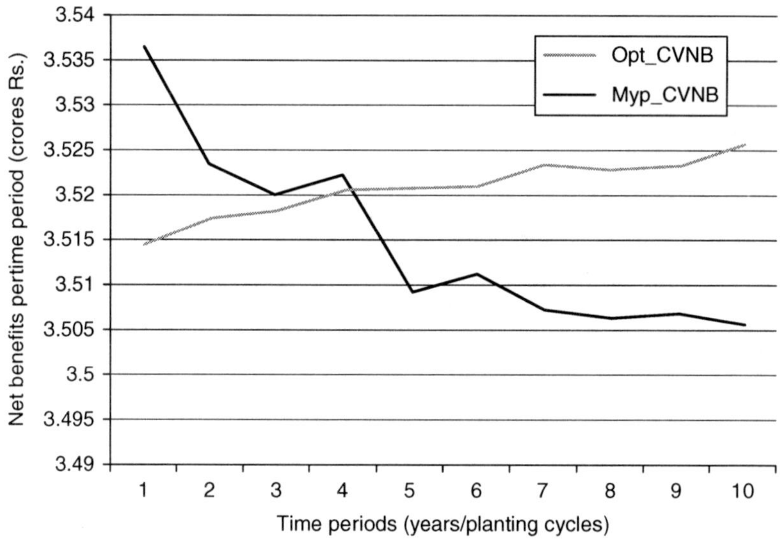

Figure 4: Comparison of current value net benefits under optimal and sub-optimal extraction (Shivni village).

If we interpret the gain in the initial period as being the payment which we would have to make to each irrigating household, in order to compensate them for curtailing their water usage, so that they behave more like the social planner, then it would require twice the payment in Hivre Bazaar than in Shivni to enact such a system. In the case of Shivni, an initial incentive payment of Rs 5000 per household might encourage them to adopt a better plan for groundwater management – such that they can avoid the future costs as a 'dividend' to better resource usage. Even though the avoided cost is almost half of what it is in Hivre Bazaar, it is still a considerable sum, and is non-trivial for many of these households, in terms of what it represents for their overall welfare.

5 Synthesis and conclusions

The choice of an appropriate instrument for water management, which minimizes the administrative and informational burden but, at the same time, maintains key water-saving incentives, represents a common trade-off that is faced by policy-makers, when implementing market-based instruments to improve the efficiency of decentralized schemes for resource allocation. The excessive information requirements of implementing an optimal tax on volumetric extraction – in terms of both monitoring the volumes of water extracted by each individual and deciding on the welfare-maximizing level across all agents – has to be balanced against the likely transaction costs that would be incurred by the individual agents who try and enact trades within a decentralized framework.

It remains an empirical matter to determine how large the loss of welfare due to transaction costs is, within a decentralized allocation scheme, relative to the administrative 'errors' that would be incurred by implementing a centralized allocation of quota with imperfect information. The advantage of allocating land areas, from an administrative point of view, lies in the fact that it is much easier to observe land area and cropping patterns, than it is to observe individual volumetric quantities of water withdrawal. So, by replacing the taxing of groundwater pumping with the restriction of land area, based on the type of crop that is grown – such that more water-consuming crops can be limited to a maximum area – we can overcome some of the information problems that would otherwise face the regulator in implementing a demand management scheme over many irrigators.

In this chapter, we have discussed the relative effectiveness of price-based instruments that increase the volumetric cost of water, compared to a quantity-based limit on available land, within the context of a profit-maximizing irrigator. We have argued for the relative efficacy of these instruments, on the basis of the transaction and administrative costs that might be incurred in their implementation, either within a centralized or within a decentralized scheme of water demand management. We conclude that quantity-based instruments are more easily observable, and might impact total consumptive use of water within a groundwater basin to a greater degree than price-based instruments that only act to reduce water withdrawals from the common pool resource.

We have also demonstrated how the value of avoided pumping costs in the future might be interpreted in terms of a dynamic problem, in which the user foregoes more immediate gain to commit to a smoother and more sustainable path of resource extraction. This kind of behaviour, however, involves moving beyond a myopic point of view and adopting a more forward-looking and patient attitude towards resource extraction and usage. The environmentally conscious village leader might be able to instil that kind of behaviour within the minds of villagers, but in the absence of such a leader, well-targeted economic instruments would need to be implemented.

Despite the welfare losses that are inevitable, when implementing market-based schemes for decentralized demand management, these might still serve to overcome problems of asymmetric information that will inevitably occur when a less-informed central administrator faces a large number of individual irrigators. The degree to which these transaction costs might reduce the efficacy below that of a 'naïve' implementation scheme by a central regulator remains a question that can only be answered by further empirical work. The importance of irrigation to Indian agriculture warrants such research, as it will undoubtedly enhance the knowledge base that policymakers need to have to better promote the sustainability of India's limited and increasingly stressed water resources.

Acknowledgement: The author gratefully acknowledge the support of Dr. Shrikant Limaye and Yugandhar Mandavkar in obtaining the village-level data for Maharashtra that the modeling results are based upon, and for the advice and project support of Drs. Stephen Foster and Sanjay Pahuja.

References

Bellman, R.E. 1957. *Dynamic Programming.* Princeton University Press: Princeton, NJ, USA.

Crook, F.W. 1999. Water use and crop production in China's Hai river basin. In *Proceedings of WCC-101, 'Chinese agriculture and the WTO'.* 2–3 December 1999, Seattle, Washington, USA.

Feinerman, E. & Knapp, K.C. 1983. Benefits from groundwater management: magnitude, sensitivity and distribution. *American Journal of Agricultural Economics,* 65: 703–710.

Kajisa, K. & Sakurai, T. 2005. Efficiency and equity in groundwater markets: the case of Madhya Pradesh, India. *Environment and Development Economics,* 10(6): 801–819.

Kumar, M.D. 2005. Impact of electricity prices and volumetric water allocation on energy and groundwater demand management: analysis from western India. *Energy Policy,* 33(1): 39–51.

Kumar, M.D. & Singh, O.P. 2001. Market instruments for demand management in the face of scarcity and overuse of water and Gujarat, western India. *Water Policy,* 3: 387–403.

Meinzen-Dick, R.S. 1996. Groundwater markets in Pakistan: Participation and productivity. *IFPRI Research Report 105*, International Food Policy Research Institute, Washington, DC.

Prakash, A. 2005. *The Dark Zone.* Orient Longman, New Delhi, pp. 75–100.

Scott, C.A. & Shah, T. 2004. Groundwater overdraft reduction through agricultural energy policy: insights from India and Mexico. *Water Resources Development,* 20(2): 149–164.